Robin Collyns was born and educated in New Zealand. He began his research on DID SPACEMEN COLONISE THE EARTH? in 1961. Early research was undertaken in London, at the British Museum, Natural History Museum, Geological Museum, Science Museum and the Victoria & Albert Museum, and in New Zealand, at the Auckland Institute and Museum, and the Gavin Gifford Memorial Museum, Te Awamutu.

Robin Collyns has had basic training in the field of electronics and has also undertaken discussions with the Buddhist monks at Colombo's Kotahena Shrine.

'Fascinating'
Evening News

Did Spacemen Colonise the Earth?

Robin Collyns

Mayflower

Granada Publishing Limited
Published in 1975 by Mayflower Books Ltd
Frogmore, St Albans, Herts AL2 2NF
Reprinted 1975, 1977

First published in Great Britain by
Pelham Books Ltd 1974

Made and printed in Great Britain by
Cox & Wyman Ltd
London, Reading and Fakenham
Set in Intertype Times

To my parents, in appreciation

Throughout this work the term 'billion' is used in the sense of a thousand millions or 10^9, and for weights and measures the metric system is used.

CONTENTS

ILLUSTRATIONS

Acknowledgement

I am grateful to the following person for his encouragement to me while researching and writing this work:

Mr. William E. Moser, BA, JP, FRAS, FRS, Astronomer; Astronomy lecturer since 1923, President of the British Astronomical Association, NSW Branch, Sydney Observatory; Vice President and Honorary General Secretary of UFOIC (UFO Investigation Centre) Sydney.

Picture Credits

Nos. 2 & 3: Ambassador College. No. 11: Zentralbibliothek, Zürich. Nos. 16, 17 & 18: John Fairfax & Sons Ltd., Sydney. No. 19: Auckland Institute & Museum. Nos. 21, 23 & 24: Mr. William E. Moser, BA, JP, FRAS, FRS of UFOIC.

'There were men from the sky in the earth in those days'

From the Judaic Kabbáláh (Heb.), *Sepher Hazzohan* – Book of Light or Zohar; written in a form of Aramaic. circa A.D. 12th–13th century.

According to Rosicrucian occult knowledge, all Kabalas – Jewish, Christian mystical, magical, alchemical, Rosicrucian and Hindu, originated from Atlantean 'mystery schools'.

INTRODUCTION

Siddhârtha Gautama the Buddha (563–483 B.C.), long ago addressed his monks, in which he stressed the dominant theme: not necessarily to believe the word of others, merely because they commanded higher authority, but to form one's own conclusions, from one's own studies, examinations and evaluation.

Why do so many inhabitants of our planet apparently possess such an inexplicable certainty or 'knowingness' that beings of high intellect and perception exist throughout the cosmos? Could this be a racial memory of times now long since gone, when the pristine 'gods' of pre-history overtly visited Earth in shining 'celestial cars'? Or perhaps there is a universal fount of knowledge and wisdom where the sum total of everything that has ever happened is inexorably recorded in the ether, and can be tapped by spiritually receptive beings?

Maxim Gorky, the Russian author, said:

> 'Generally speaking, there is nothing fantastic in life. All that seems mysterious has a very definite basis in reality ... There is nothing that man has thought up that does not have roots in the real world.'

Aside from metaphysical questions, maybe there is a more 'down-to-Earth' (literal) reason why Earthman has an intuitive awareness that others similar to himself exist on distant worlds?

Research into many very old and authoritative records of Earth's chequered history indicates an overwhelming probability that our world has, on countless occasions, been host to extra-terrestrial spaceship visitations.

Where do the spacecraft originate? It would of course be more than difficult to attempt to identify with certainty from which planets or solar systems they come. Astronomical estimates indicate 100–200 billion stars in our own Milky Way galaxy; approximately 60–67% of these are considered to possess planetary companions, while upwards of two to six billion galaxies exist in the universe. By comparison Earth is but a grain of sand. Throughout the vastness of space there must be – according to the laws of mathematical probability – planets both older and younger than Earth. Some eight billion planets in the Milky Way have been theorized to be possibly inhabited. Is it not impossible to believe that some, at least, of all planets, support sentient beings who aeons ago solved the problems of space travel, as undoubtedly we will before too long?

As recently as 1965, world-famous scientist Dr. T. E. W. Schuman, Deputy Chairman of South Africa's Atomic Energy Board, said:

> 'During the present century no human being, Russian or American, will land on the Moon and return safely to Earth.'

We have progressed far since then, with manned Moon flights now taken for granted by the public.

Konstantin Eduardovitch Tsiolkovsky (1857–1935), the Russian founder of 'cosmautics', cherished positive convictions on the reality of space flight, and a trust in Man's abilities. A stone obelisk in Kaluga is engraved with this famous quote from Tsiolkovsky:

> 'Mankind will not remain bound to the earth forever, but in the pursuit of the world and space, will at first timidly penetrate beyond the limits of the atmosphere, and then will conquer all the space around the sun.'

In this present era the prophetic words of Tsiolkovsky are beginning to prove true, but have others also timidly penetrated beyond the limits of their atmospheres and then conquered all the space around many suns?

The Mediaeval and Renaissance fallacy of Earth as the sole planet supporting life is quite outdated. World opinion has greatly changed since the days when Giordano Bruno, a sixteenth-century Italian monk, was convicted of heresy for daring to forward the outrageous suggestion of other inhabited worlds in the universe. Bruno's punishment for this supposedly heinous crime, was to be burnt at the stake on the Square of Flowers in Rome, 17th February, 1600.

If the Theory of Evolution – which is really an unproven faith – is reflected upon, it could be proposed that Earthman is the sovereign expression of an organism unfolded to the most elevated status in the evolution 'chain'. Accordingly, on planets with similar atmospheric and gravitational conditions to those on Earth, similar evolutionary results may prevail, or, 'parallel evolution to a cosmic scale'. But *did* Man evolve?

For those curious as to why spacemen do not initiate public meetings with Earthlings, it may be relevant, and/or related, to note the context of a report prepared by the Brookings Institution, Washington D.C., for the National Aeronautics and Space Administration (NASA) in 1960, which in part comments that:

> 'An all-out attempt to contact intelligent beings on other planets could lead to sweeping changes, perhaps even the downfall of civilization itself.
>
> 'Even on Earth, societies sure of their own place have disintegrated when confronted by a superior society, and others have survived, though changed.
>
> 'Clearly, the better we can understand the factors involved in responding to such crises, the better prepared we may be.'

One indication of life throughout the cosmos is that 1% of meteorites analysed, exhibit carbon atoms – essential to all life-forms – linked in a characteristic structure, indicating that once they formed part of an animal or a plant. A small percentage of meteorites dated at 4.5 billion years, contain in their crystal structure the chemical compound 'sporopollenin'; sporopollenin contains spores once part of advanced life-forms, fungi, and plants.

Before looking at the probability of spacecraft visitations to our planet Terra in vast antiquity, it is relevant to this work to know two new discoveries about Earth in our relation to the universe.

THE REAL AGE OF THE EARTH

Four and one half billion years is the accepted age for our planet, but recently, E. K. Gerling, a Soviet scientist, analysed rock formations from deep beneath the Baltic Crust which gave a *consistently* advanced age to 6.5 billion years. Both dates were estimated by measuring radioactive Potassium-40 as it changes to Argon-40, but the rock samples from beneath the Baltic Crust were at a considerably lower level than previous samples collected by scientists.

Gerling now believes Earth's age may exceed 11 billion years: if proven this would, of course, cause upheaval in scientific thought relating to the age of our world (and Earth could last for a further 6 billion years).

THE FIRST LIFE-FORM ON EARTH

This was blue-green algae, fossilized plant cells of which were discovered in Africa, and dated at 3.2 billion years.

UNIVERSAL LIFE

Dr. Jan Gadomski, the eminent Polish astronomer, said:

> 'Highly developed civilizations on planets of the suns in our galaxy, should be the rule, rather than the exception.'

Dr. Lewis W. Beck, University of Rochester, Mass., remarked that:

> 'There are many abodes of intelligent life in the universe, and many of them are inhabited by organisms with intelligence far higher than ours.'

SPACEMEN

Innumerable old codices, legends, etc., unmistakably refer to spaceships – 'winged wagons', Icelandic legends – and spacemen; the Brahmaputras and Manasaputras (Sanskrit); B'ne-aleim (Judaic); Lords of Venus and

Mercury (Rosicrucian texts); Sons of the Sun (Inca legends), and many others.

A 1970 archaeology discovery and mystery, was fifty male human skeletons, in a 'mass grave' dating back to the days of Imperial Rome, located near picturesque Terracina, 96½ km south-east of Rome.

All the skeletons – not African slaves – are taller than two metres, which contrasts sharply with the average height of Earthmen at that time, who were considerably smaller than the average height of Earthman today!

The bodies were buried without any identification or possessions of any kind, in plain, tiled coffins. Burial without possessions was also unusual for those days.

At that time, the life expectancy was 22 years, yet all these unidentified men – evidently in perfect health and aged between 30–40 years – died together. Why, the archaeologists wondered, and who were they? Were they the skeletal remains of Roman gods or spacemen? Or spacemen whose UFO may have crash-landed and punctured, thereby depressurizing the craft and inducing fatal 'bends' in the crew by lethal accumulation of nitrogen in the nervous system? Could this explain the mystery men who all died together, apparently in perfect health?

For many years UFO authors have theorized that extra-terrestrial spaceships are propelled by magnetic forces. In 1970, Drs. John Allen and Geoffrey Endean, two Oxford University scientists, announced that electromagnetic fields in the Crab Nebula more than 6,000 light years away were, it seemed, travelling at *twice* the speed of light. Einstein's special Theory of Relativity postulated that the speed of light was the fastest movement in nature, but this new discovery could indicate that magnetic, or electromagnetically propelled spaceships from alien planets, bridge the enormous distances between the stars in *comparatively short* time periods!

The Hindu chronology refers to a time some 4,000,000 years ago during Sat-Yuga ('Golden Age' – 'The Age of Truth') 'when the souls now on Earth first arrived here'. Were our ancestors brought to this sphere in spacecraft? In this context there may be a link with the 'Castle', a startling massive rock formation in Argentina's Quebrada Canyon, which bears striking resemblance to a Cosmodrome artificially shaped in antiquity. The Castle is said to bear a closer resemblance to human-created architecture than any similar formation in the world.

Still another Cosmodrome for UFOs seems to be the Great Sand Dunes in the Sangre de Cristo ('Blood of Christ') mountain range in south-central Colorado. This area, covering 146 km^2 and 2.7 km above sea-level, is the site for frequent reports of UFO landings.

What of extra-terrestrial life in our own solar system? In late 1971 Soviet astronomers in Azerbaijan announced that spectra had been obtained from the dark side of Venus for the second time. The carbon dioxide content of the Venusian atmosphere prompted head astronomer Nadir Ibrahimov of the investigating team to comment:

> 'Seeing that all life as we know it is based on carbon, we cannot rule out the possibility of some form of life on the planet.'

Life is exceedingly tenacious, and can survive and thrive in the most adverse conditions. Exobiologist Dr. L. Lozina-Lozinsky of the USSR, working with two other exobiologists (biologists who unravel the mysteries relating to extra-terrestrial life), discovered that the protozoan micro-organism Colpoda maupasi, which lives in soil, could vigorously flourish in a *carbon dioxide* atmosphere with less than 1% oxygen, at pressures only 1/12th to 1/50th of those found on Earth. Even more striking was the revelation that the Colpoda survived in daily temperature changes from +5° to −30°C.

Throughout this work many enigmas associated with spacemen of antiquity are examined. For example, why did US electron microscope observations reveal a *pathological excess* of calcium in the shells of dinosaur eggs from the late Cretaceous period? As many as seven layers of shell were found in some eggs! Stress effecting malfunction of the pituitary gland will cause excess calcium to be formed in some birds' and turtles' eggs, but there is an *alternative* – and more logical – reason why too much calcium built up in the dinosaur eggs, and that is *genetic mutation*! *What* could induce genetic mutation? Intense radiation from a Supernova, or intense radiation from a *terrestrial* cause? Whatever the reason – investigated in Chapter 5 – the dinosaurs disappeared very (comparatively) quickly. Remaining dinosaurs bearing eggs later would not understand the reason why their babies could no longer break through the shells. Something *violent* happened which destroyed the dinosaurs in huge numbers, then the remaining reptiles gradually died out, ostensibly as a direct result of the earlier catastrophe!

Critics of UFOlogy say there is little or no evidence for extra-terrestrial spacecraft visits to Earth in historic or earlier times, but during research for this work I found the opposite to be true: i.e. a plethora of early spaceman/UFO data which has, due to its volume, necessitated careful selection for the manuscript.

It has become increasingly evident to thinking people that neither the theological doctrine of 'divine creation', nor the Theory of Evolution, can explain the origin of Man, flora and fauna, on this or any other planet; noticeable weaknesses and inconsistencies are apparent in both beliefs. In this work we examine a third possibility for the origin of life-form bodies, based on my own twelve years of research into rare philosophic treatises and discourses, old lexicons and codices, very advanced spiritual Yoga and Arabic spiritual theology teachings, and *many* other sources, which in this context includes physical/material discoveries to support this concept.

Space does not permit covering every subject in detail in this work. The following briefly mentioned curiosities are only three among countless other religious and non-religious aerial phenomena which could be further researched in the UFO field.

For example, the secret of Fátima which was not released to the world by the Vatican in 1960 after a specific earlier request that it should have been released in that year. Was the secret that other planets are inhabited by people like ourselves? A religious aerial phenomenon, similar to the 1917 Fátima Portugal mystery, is regularly seen from the sacred Wu'Tai mountain in northern China but this particular aerial phenomenon is associated with Buddhism. Then there is the 'phantasm' – 'lady in white' – still another UFO-like aerial manifestation, first seen in 1968 over the cupolas of Coptic churches in Cairo.

This Introduction covers but a partial cross-section of the type of material contained in forthcoming chapters. If at first it appears that unrelated diverse subjects are covered in Chapter 1 this is not so, for all are component units of the whole and are concisely coalesced in an end-of-chapter summary.

Similarly, the first part of Chapter 2 seems to regress still further back into pre-history instead of being contained in Chapter 1, but this too is a planned step, for by the end of this book the reader will find that these and other variations do follow a plan of gradual subtle progression on a gradient scale from hoary pre-history to our present twentieth century.

This is a *different* kind of history book . . .

Robin Collyns,
Auckland, New Zealand, 1973.

1. DID EARTHMAN'S ANCESTORS TRAVEL 26 LIGHT YEARS FROM LYRA?

'I have chosen that part of philosophy which is most like to excite curiosity; for what can more concern us, than to know how this world which we inhabit is made; and whether there be any other worlds like it, which are also inhabited as this is?'

(Bernard de Fontenelle (1657–1757). Preface to: *Discourses On The Plurality of Worlds*, 1686.)

Before describing 'traces' left by spacecraft visits to Earth in archaic times, it is first necessary to journey far, far back through time; back to a period so remote, that in this philosophical and metaphysical study we will look at possible reasons for the origin of Man, how Man first appeared, how long Man has dwelt on Earth, and whence he came. Is Man's memory occluded – or has he lost remembrance of his advent upon Earth? We may not merely be observers of the UFO phenomenon, for whether we realize it or not, we may be involved.

In *The Path Of The Soul* (Shabd Yoga – Divine Inner Sound), a curious work published in the holy city Amritsar, is this dramatic passage:

> 'The length of time that has passed since the soul entered the physical universe is virtually incalculable, for the age of the human race defies imagination. For back, millions and millions of years before the beginning of recorded history, there were great civilizations of which there now remains no trace.'

Is the human race really 'millions and millions' of years old? On Earth – other planets? Did physical bodies 'evolve'? Or perhaps another possibility is that the physical bodies of humans, animals, birds, insects, microbes, etc., were collectively designed and mocked-up, and/or created, by a race of superior spiritual beings (perhaps our race?) in deep space countless millions of years ago? Did all creatures once exist only in spirit form?

The two generally accepted explanations for the origin of Man and creatures have obvious weaknesses in their hypotheses. As one example, the theological doctrine of perfect momentary 'divine creation' does not easily reconcile with observed characteristics of the human body which is, it seems, subject to too many structural and biochemical frailties and imperfections to have first appeared in this way. The vulnerability of the human body to disease epidemics and to impact blows, are just two examples among many bodily faults. The Theory of Evolution, the other explanation for the origin of Man and creatures, also has weak points – the missing link is one of the most notable. Also, *no* creature today is observed in an intermediate transitional evolutionary stage. This is *not* to be confused with slight bodily variation of creatures *within their own species* as environmental adaptability characteristics.

Indeed, many species of fauna and flora surviving today are essentially identical in appearance and bodily structure with their fossil ancestors up to hundreds of millions of years old; they have not 'evolved' or changed form – to the perplexity and dismay of evolutionists – but have remained the same, or in some cases with minimal environmental adaptability variations, since their earliest fossil records on this planet. When were they first brought to Earth in spaceships? According to the evolution doctrine, the great age of the earliest fossil species mentioned here is considered far too long a period for life-form bodies to have remained in a non-flux, non-evolutionary static state!

Included in the category of many creatures basically unchanged since their earliest fossil records – as far back as 550–570 million years – and excluding flora, the following partial list gives the closest approximate ages of the oldest fossil ancestors discovered,

Coelacanth (fish) — 350 million years
Neopilina galathaea (deep-sea mollusc) — 350 million years
Cockroach — 350 million years
Dragonfly — 170 million years
Tuatara lizard (NZ) — 135 million years
Black ant (Formica fusca) — 70 million years

Man . . . Judaeo/Christian and other theologians, and evolutionists, misguidedly or unwittingly endeavour to denigrate and self-depreciate/self-degrade Man from the noble, elevated spiritual being that he really is. Theologians say Man was created in the spiritual image of God, yet theologians regard Man as a miserable sinner – an almost unworthy entity – which also reflects on God, as being subject or falling to, a less than perfect state. Evolutionists disparage Man by associating him with a primitive animal origin, which to many sensitive thinking people is an abhorrent thought. The Theory of Evolution is just that – a theory – it is *not* proven fact. Modern Man bears but a vague superficial resemblance to an ape; anatomically, the central nervous systems of Man and ape differ. Even with complex cerebral, etc., structures, an ape still cannot be taught to speak in human vocal language. Clearly, both concepts for the origin of Man cannot be acceptable to many free-thinking beings who believe every man to be a sublime free spirit.

Man was/is not an animal. Soviet researchers have shown that Man's aggressive tendencies are not biologically inherent in his genes – a carry-over from theorized omnivorous animal ancestors – but are tendencies conditioned into his thought patterns by society. Hostility in Man to the point of irrational psychological motivations is *not* instinctive.

On another note, sages and savants of old have learned that Man's special unique creative abilities, aside from procreation, are not due to diverted sexual energy as popularly speculated, but originate solely from his personal spiritual aspirations. By nature, Man is a creative spiritual being, constantly newly-creating and conquering new horizons, possibly on all planes of existence.

In this work it is to be hoped that the reader will consider the evidence to indicate that spiritual Man is (basically) an intense God-like being who has lived in his present bodily form for possibly hundreds or thousands of millions of years, *not* just the few thousand years since Cro-Magnon Man appeared in Europe.

Did Man suddenly appear on Earth in his present 'modern' form? Perhaps 'evolution' took place on other planets before modern Earthman mysteriously appeared or arrived – by spaceships?

Certain discoveries indicate a sudden arrival of modern Earthman who *was not* related to primitive primates, apes or hominoids.*

If it was discovered that on Earth in antiquity *modern* Man actually co-existed with early primates, apes and hominoids, this would of course indicate a definite non-relationship between modern Man and these creatures. That is, (all?) these species separate from each other, which on this basis would *also* rule out the hypothetical evolutionary pattern paralleled on other worlds.

Modern Man today coexists with the non-human anthropoid ape families. How could he have had any evolutionary 'link' with any of them, or with any other surviving anthropoid primates? The apes, etc., should have disappeared. It does seem illogical to assume that some of their ancestors strangely 'branched out' to eventually become men, with others manifest as they are today.

If modern Earthman really did have, or has, a 'link' with the non-human anthropoid primate families, then he should be able to cross-breed with them, but of course this is impossible. The DNA – deoxyribonucleic acid – molecules within living cells have, for every life-form body, such a rigid set of different coded 'blueprints' for cell reproduction that (unrelated) species cannot crossbreed.

*Hominoids, from Lat. 'Homo', Man, and 'oid', alike; i.e. for the purposes of this book the term Hominoids excludes apes and refers specifically to the man-like 'Ape Men'.

Also, as a hypothetical question, if several thousands of millions of years ago before the physical universe began, all types of countless trillions of beings existed only in spiritual form or in a spiritual state, perhaps a point was reached when, for example, the most superior spiritual beings (our spiritual race, or very advanced human souls?) desired or needed to live in physical bodies in a physical universe. Was it a higher force – the Supreme Creator – or perhaps, collectively, this advanced race or races of spiritual beings *themselves* who produced the power – radiant energy – to create mass, physical bodies for themselves and possibly creatures? Did such beings *also* create the physical universe? How did the physical universe come into existence?

Was it possible that this hypothetical superior race or races existed as a knowing-all static in pure spirit form, so created material mass – worlds, bodies – in order to occupy the aeons by constantly conquering new horizons in the physical universe? Alternatively, the (temporary?) reason to live in a physical universe may be necessary to gain new experiences to further advance the soul. It seems doubtful though, that 'new souls' are constantly being created, as has been theorized.

Many of the concepts outlined in this book may seem extraordinary, but science which requires material visual/audible, etc., evidence or proof on the physical plane, cannot explain certain subjects which cannot even be explained by orthodox theology. How, for example, could science or theology explain the origin of something that did not have a beginning – something that *always* existed?

The Sanskrit teachings of the 'heavenly region' called by the ancients 'Sat Desh' or True Home, in which spiritual Man existed, long before he possessed a chemical body and which was also the abode of the Supreme Creator, indicate among other sources that *spiritual Man is immortal, without beginning or end*.

If the soul of Man always existed, then 'old' or 'new' souls is an incorrect belief among those who adhere to this thought. As a very elementary example of this, a highly advanced spiritual Yogi is obviously far ahead in spiritual development compared with a still Stone-Age New Guinea tribesman, who in the hinterlands and jungles exhibits little or no progression since antiquity. But the Yogi has not gained his powers and wisdom by existing as an older spiritual being, but by expressing as a superior soul.

Every human psyche is unique – immeasurably powerful and individualistic and very different from every other being in personality, aspirations, sensitivity or lack of it, etc. After thousands of years the New Guinea jungle tribesmen have not progressed essentially from the Stone Age; higher wisdom, sensitivity, and so on have not come with time, but the potential for greater spiritual development must be inherent in all men. By contact with advanced peoples, and by supervision, the New Guinea tribesmen achieve a workable civilization. At their present stage though, none (and few among ourselves) possess the spiritual abilities to enable them to undertake a journey to the Himālayas to take their place alongside eminent Yogis, who can reputedly teleport their bodies, materialize flowers and fruit, etc., from the ether, and perform many other feats.

Why would spiritual beings be more, or less, advanced? Are we all in varying stages of spiritual unfoldment or development, in which eventual spiritual perfection is achieved by all? Or do some remain static, never progress, or even regress?

Even in our own advanced but largely materialistic Western civilization which dates back at least several thousands of years, few souls have spiritual powers comparable with Himālayan Yogis; notable exceptions were Luther Burbank and Therese Neumann, but many souls do have undeveloped latent powers of spiritual healing, ESP, telepathy, skin vision, water divining, etc., which could be further developed. So the potential for spiritual development to a higher level does exist among many, most, or all souls, undoubtedly primitive tribesmen also. Perhaps a tribesman from the highlands of New Guinea could, one day, after many rebirth experiences among more advanced peoples – peoples with superior, physically inherited genetic, brain (computer) characteristics – take his place alongside Himālayan Yogis.

If 'new souls' are not constantly being created, how could the phenomenal population increase be explained? If, looking at the reincarnation enigma, all human souls were reborn immediately after the demise of their physical bodies, this still could not entirely explain the vast increase, for the population of this planet is expanding at a constantly increasing rate. If all these new people are not new souls, then perhaps souls from other and/or less advanced planets are reincarnating on Earth, possibly as a step forward or temporary regression for erring space beings, in spiritual development in a possible galactic or universal planetary cycle. Perhaps, too, advanced space beings are reincarnating on Earth, or extra-terrestrials may even arrive physically in spaceships!

The fourth world UFO congress held at Wiesbaden, West Germany, was informed in 1960 that a census revealed that 800,000 people from other worlds were at that time living on Earth. (But how was the census compiled?)

Another consideration for the enigma of increased population is the possibility of souls dwelling in physical bodies for the first time, leaving the spirit dimension (Heaven?), as in the possible case of primitive tribes, which may be stepping stones – or even regressive steps for Hitlers, etc.?

It seems unlikely that the (sexless?) spirit of Man started his first bodily experience in the form of a microorganism; several scientists believe that Earth is not sufficiently ancient for intelligent beings to have 'evolved' from protozoa, which are microscopic unicellular organisms. From whence did they come? The Biblical *Genesis* 'Creation' account is too allegorical to be a source of serious data for the creation of life-form bodies, so where do we find the evidence? The fossil record, though extremely scarce in this respect, but in conjunction with ancient Eastern philosophic and related treatises, indicates that separate – not hermaphrodite or androgynous, bi-sexual – men and women, have existed in their present bodily form on other planets for at least several hundred million years, before migrating to Earth more recently. By spacecraft?

Could spiritual Man collectively create a bodily form for himself solely by radiant spiritual energy? The creation of physical/material/chemical matter by radiant spiritual energy is a subject known and utilized by exceptionally advanced Himālayan and other Yogis. Of possible importance in this connection is the historic Winnebago Red Indian 'Creation' legend which in obscure fashion says:

> 'As everything happens just as I wish it, I shall make a man like myself in appearance.'
>
> (From *The Many Worlds of Man*, by Jack Conrad.)

Did the spirit of Man design his own bodily form? Duplicate bodily parts may indicate collective intelligent planning; e.g. two ears, eyes, lungs, kidneys, adrenal, endocrine and reproductive glands, etc. If these indications really are of collective intellective design in the human, for example, body, were all duplicate components and other factors originally created by spiritual energy radiant from group souls, if this was so, or were some a later refinement (by also bi-sexual?) scientists in laboratories on other planets? That is, and/or part of a genetic alteration/separation from an earlier hermaphrodite/androgynous self-reproducing bodily form, of which apparently, once hermaphrodite/androgynous remnant indications still remain in Earthman; viz. mammary gland vestiges in the human male, which may relate with an allegorical account in the Biblical *Genesis* evidently referring to this very concept. By this, I mean *Gen. 1:27* which states: 'Male and female created he* them' which indicates a *first* hermaphrodite/androgynous bodily form for Man; then the genetic separation: – 'she shall be called Woman, because she was taken out of Man' (*Gen. 2:23*) ... by spacemen scientists? Or, in simple terms, did scientists separate bi-sexual Man into

*God.

male and female? This is the confused 'two creation' account believed in by some theologians.

If Earthman did arrive from other planets, his basic racial characteristics could be explained within my hypothesis: e.g. hereditary factors pre-programmed into his genes in extra-terrestrial laboratories as survival/adaptability agents for differing planetary environments; example: Chinese, Negroes, Caucasians (and others?) We may have been more ideally suited to atmosphere, gravitation, etc., on our home planets – if in fact our ancestors did come from other planets – and possibly longer lived, but structural/chemical bodily weaknesses are still evident in Man.

Some animals and even plants may also have been genetically separated from an hermaphrodite/androgynous bodily form by space agencies. But *if* all this really was so, then why the separation? To coerce spiritual communication, and thereby also self-improvement, by seeking other beings? And were 'instincts' also programmed into animal genes by spacemen scientists? For example, U.S. scientists discovered that birds are attracted to the colour yellow, which seen from the air is the natural colour of cereal crops!

Referring again to the aforementioned duplicate bodily parts, if necessary Man can survive with only one each of such organs although in a reduced capacity, but still alive and reproducing. Obviously, all these bodily parts were intended to function at optimum efficiency by working as a pair, but if only one super-organ existed instead, performing the work of two, e.g. a median Cyclops eye with a 160° field of vision, then damage to, or loss of that eye or organ, would mean disaster for that particular function and in some cases even death.

Are duplicate bodily parts emergency spares? Do we possess but one heart due to an incompletely developed body? In the situation of the loss of a duplicate limb or sensory organ (eyes, etc.) the theory of other limbs and sensory organs (ears, etc.) becoming stronger or more acutely sensitive to compensate may, in this context, also apply to the heart which is the strongest muscle and organ in the body. Similar observations could apply to other life-forms.

It does not seem possible for bodily forms to have 'evolved'! Without (for example) all the component parts of an ear: – pinna, meatus, drum, hammer, anvil, stirrup, etc., or of an eye: – optic nerve, retina, choria, sclerotic, cornea, crystalline lens, pupil, iris, vitreous and aqueous humour, *all* functioning together, complementary and in perfect precise harmony, and in harmony with all other bodily parts, then hearing or vision could not take place! All the component parts of an ear or eye must have appeared at the *same time* for these organs to be able to function! How could slow 'evolution' from protozoa explain this? Perfect instantaneous bodily creation is not indicated either, as structural and chemical bodily weaknesses are evident in Man (and animals?) in addition to vestigial organ remnants of a once bi-sexual body.

The extremely precise balance mechanism in the skull, the three U tubes filled with fluid, each in a different plane, like the previous observations applying to the ear and eye, also indicates intelligent design of a mechanism for *erect* mobile Man, *not* slow 'evolution' from protozoa!

Man's molar wisdom teeth which erupt about the twentieth year, long after the rest of the secondary teeth, although often chalky, may have originally been intended to replace in a small way, teeth which might have been lost through the years. Teeth and other nerves, which warn by pain of decay, damage, etc., may have been a built-in safety factor carefully planned. Another protective agent could be the ability of the human body to synthesize vitamin K, the blood-coagulating vitamin, by bacterial process. Carnivorous animals possess the ability to synthesize vitamin C, but Man with an intestinal tract designed (?) largely for a diet of fruit and nuts, does not have the vitamin C synthesizing ability.

Two other possible built-in protective factors are (1) the 'blood/brain barrier' which is an indiscernible biochemical 'fence' to prevent toxic substances from entering and damaging the central nervous system, and (2) the ability of the brain to produce the chemical 'scotophobin' to preserve physical memory of the spiritual fear of darkness. The cerebral cortex convolutions which compress a capacious external region containing billions of nerve cells into a compact mass, indicates intelligent design.

Another suggestion of intellective planning in the human body and in other living organisms, is the discovery by Russian exobiologists of not one, but a whole series of biological 'clocks' controlled by a central 'clock', which in turn is regulated by the subthalmic region of the brain.

The extraordinary 'pacemaker' cells in the heart's Sinus Node, whose vocation in life is to keep the heart pulsing at the correct rhythm, also signifies inventive purpose. Were the DNA molecules in these special cells (not regular nerve cells) originally pre-programmed to perform their specific function or duties in Man and animals, by extra-terrestrial scientists, or looking at the question from a philosophical point-of-view, does each bodily cell possess its own omnipotent energy and/or intelligence – supernal consciousness – or is it divinely inhabited by simple elemental spirits subservient to higher life-forms?

Professor Geoffrey Burnstock, a London-born zoologist at Melbourne University, announced that he had discovered a 'self-governing' third nervous system in the body which *does not* depend on the brain for its functions.

Still another implication of metaphysical determinism in perspicacious design – but in this case only in higher organisms – is the plethora of duplicated 'left-over' redundant DNA in genomes. Only one DNA 'blueprint' is required, but curiously is duplicated up to one million times. This appears to be a programmed factor to safeguard against faulty or mutated genes.

In Man the large arteries in the arms and legs run close to the long bones in the limbs for protection, but even going far, far back into time, it seems evident that a similar intelligent design existed in the bodily structures of, for example, dinosaurs.

The Diplodocus, the enormous reptile in the order Sauropoda of the dinosauria, possessed a very unusual arrangement of chevron zigzag bones underneath the caudal vertebrae, which protected the tail blood vessels while it was dragged along the ground.

These dinosaurs were the proud owners of two brains – one at each extremity of their bodily length. Their length was such (24 m) that nerve impulses from a solitary cerebral centre would take too long to travel the full length of the body. *Who* designed the bodily forms of these creatures?

Man's skeletal frame is, it seems, designed to protect certain organs – the brow to protect eyes, the rib-cage to protect lungs, heart, large arteries, and veins – but what may be an imperfection is the cranial shape, which if spherical, while not being the most aesthetically pleasing shape, would mathematically be a more ideal shape to protect the brain; or, equal pressure on all points, which would lessen the chances of cranium fracture.

Perhaps some spacemen's and our bodily form is one, which may in the distant future, be gradually genetically 'phased-out' in favour of a superior improved model. By concerned, spiritually and scientifically advanced spacemen? The concept of 'neo' or new humans has already been considered by U.S. scientists as a feasible technological goal.

Venous detour and subsidiary nerve channel rehabilitation abilities in Man's body indicates intelligent design, but what might be the discovery of a strange katabolistic or breaking-down factor, was established only in 1970 by researchers into the study of ageing processes.

Does the human, and animal, etc., body have a built-in obsolescence or 'self-destruct' factor, or is the following, another bodily imperfection or weakness?

The mineral iron is essential for health and without it Man's body would expire, but a team of Ukrainian scientists at the Soviet Academy of Sciences' Institute of Gerontology and Experimental Pathology announced that experimental research indicated that the DNA molecules which carry hereditary data in the body and muscle protein molecules, *age* – become brittle and less durable due to increased iron content in the DNA molecules. Here indeed is a mystery. Man's body cannot function without iron, yet ostensibly, iron is indicated as an ageing factor.

Let us now look at another enigma – the origin of the physical/material universe. The 'steady state' 'big bang' 'oscillating theory' 'anti-matter' concepts do not and cannot explain the beginning of the physical universe! The first element required for the beginning of the physical universe is hydrogen gas, but *from whence* did the *first* material hydrogen atoms or atom come? *Science doesn't know! Hydrogen came into existence out of nothing!* There is much material for deep thought in that statement.

Astronomical observations indicate that (this) physical universe did have a beginning – a starting point – about 15 to 20 billion years ago, but how? The physical universe, and life-form bodies, it seems, had definite starting points, but before that there apparently existed spiritual beings – (the directive intelligence behind it all?). Can even the 3,000 or so tonnes of cosmic dust which daily rain upon Earth be part of a pre-ordained pattern for partial physical replenishment of a planet?

And does the physical universe end? (35 billion light years in diameter?) Is there a point where galaxies, suns, planets, etc., thin-out in space until no more can be seen? And what might lie beyond that – *infinite emptiness or nothingness?* It is very difficult, and even disturbing for some people, to attempt to grasp this thought. Even a constantly expanding physical universe must also expand into an infinite nothingness of empty space! Is there no end? Empty nothingness (may) even be in the process of constant creation or expansion into *never-ending* infinity, or may pass into a different and unknown dimension or dimensions, as constant flux or change is an observed characteristic of everything in the universe. But perhaps infinite emptiness or nothingness remains static – it always was, and always will be, without a beginning or an end? We on Earth who are taught physics from material/physical laws find it incomprehensible to begin to understand such subjects. It is above our limited understanding to relate with concepts outside our frame of reference.

Perhaps spiritual beings, biochemical bodies, and the physical universe were 'created' by the Supreme Intelligence (God), but did spiritual beings always exist, or did they come into existence from nothing? Archaical Eastern, spiritual, and other teachings, indicate that spiritual beings *always* existed. And are each of us, though *individual*, a component fraction of the Supreme Creator? Is God, in part, the omnipresent radiant energy (the *first* force required?) from *all* spiritual beings collectively in the universe who *always* existed?

A particular branch of Súfism, the Arabic spiritual theology – Tasawwuf – teaches that the physical/material universe is an 'illusion'. Possibly, collectively, spiritual beings could, if they created the physical universe by radiant energy – or instigated the beginning – also affect the material universe to implode and disappear, thereby reverting to the original pre-physical/material state? It is significant to note in this connection the documented reports of spiritual Yogis who can, on a small scale, reputedly materialize flowers and fruit, etc., from the ether, through their personal motive spiritual energy radiant from their own divine consciousness.

Diverse interpretations have been accorded the following quote, but does it *really* mean that spiritual Man in deep space created for himself a physical body? Note the similarity to the Winnebago 'Creation' legend:

> 'And God said, Let us make man in our image, after our likeness. . . .'
>
> *Gen. 1:26.*

'us' – 'our' – a multiplicity of gods?

Remember, the Sanskrit knowledge of Sat Desh teaches that in the pre-physical void spiritual Man originally resided in the abode of the Supreme Creator.

All creatures *must* be immortal spiritual beings living in physical bodies. They *must* be more than just bone phosphate and chemical bodily forms animated by weak electric currents generated in the cerebro-spinal fluid, or by some other material explanation. Some believe that even the humble rock is possessed of an individual (elemental) spirit, while notably, the tiniest insect will indicate a unique personality, proving that the lesser creatures do not belong to a 'group soul' as believed by some, but probably group souls.

Research into plant metaphysics proves that plants *will not survive* without acceptance from other plants and life-forms including Man, in the immediate environment. Were weeds once advanced plants, now rejected by other plants, animals and Man?

Pravda, the Russian newspaper, reported that roses and other plants have memories. U.S.S.R. research discovered that the rose releases 'electrical pulses' similar to the 'nervous pulses' of Man, which pass 'signals' to a 'centre' that processes the data and gives a 'command', then the rose 'reacts accordingly'.

Other Soviet research proves that honey bees with only a tiny brain have the ability to count. Later in this book we will see that the honey bee may have been brought from another planet. Previous thought was that only a few of the higher mammals with large developed brains could count, but the little bee can count, whereas other animals with large brains cannot.

The Russian scientists taught the bee that food could be found on a triangle but not on a quadrilateral. It was then learned that whatever the exact triangle shape, size or position, and in the presence of any four-sided figure – both the triangle and four-sided figure manipulated into many positions – the bee always recognized and went to the triangle.

The next group of experiments taught the bee that food could be found on shapes painted with two colours, but not with one or three colours. The bee always recognized and went to the two-coloured figure. The third group of experiments – the bee's final graduation test as a mathematician – was the presentation of charts with different numbers of circles. The bee always identified a three-ring chart from a one-, two- or four-ring chart, whatever the sizes or positions of the circles on each chart.

Evidence suggests that bees also exhibit Extra Sensory Perception, while it is suspected that ants have telepathic abilities, even though ants do not possess large higher-developed brains like humans, which indicates that the spirit is the primary factor in telepathy. But in the observation of people (and ants?) the physical brain may possibly be utilized by the spirit as transmitter and/or receiver. And what mysterious sense, perception, and reason, causes the Australian white ants (termites) to build their 6-m-tall earthen nest with the sharp edge always in line with the magnetic North?

Perhaps elemental rock, tree, plant, amoeba and virus spirits, etc., (of which amoebas and viruses and dwellers in the microcosm or 'small universe' which *must* range down to as small as the miniature solar systems, i.e. suns – atom nuclei, protons, neutrons – and worlds – electrons) may change spiritually to higher advanced bodily forms or amoebas, etc., as they progress? *Who* designed or created these life-form bodies?

As it appears that Man is the most advanced spiritual being in the universe (reputed sightings of spacemen report similar bodily structures), was it Man who mocked-up and/or created bodies for lesser spirits? Research undertaken by specialized investigators into animal and plant mysteries indicate that spiritual dwellers in these bodily forms are not, and will not, be the same as Man. Evidently the soul of an animal is more advanced than the elemental spirit of a plant, but it seems most unlikely that a cat could ever become a king. The soul of a cat may progress within its own sphere by rebirth experiences in animal bodies. *Genesis 1: 26* clearly states the superiority of Man over the animal kingdom:

'. . . let them* have dominion over the fish of the sea,

*Man.

and over the fowl of the air, and over the cattle, and over all the earth, and over every creeping thing that creepeth upon the earth.'

Only Man, of all beings, will devote a lifetime in spiritual pursuits to become more attuned with the God-power: – Monks, Sadhus, Swamis, Yogis, Fakirs, Súfis, etc. No animal would ever purposefully, or have the ability or even concept to, deliberately devote a lifetime to spiritually advancing its soul, and only Man, aware of the goal-achieving limitations of his life-span, is preoccupied with time.

It is Man, of all beings, who is the only entity to cherish ideals, an ability unique, although creativity well advanced in Man does exist at a simple level for certain creatures. The Australian Bower bird adorns his bower with feathers, fragments of coloured glass, etc. This appears to be in conjunction with species perpetuation instincts, but only Man has the ability to visualize, plan, and systematically carry through advanced future goals – sometimes years in the future. Short-term animal goals are possibly for survival, although when a porpoise, and in one case a seal, nudges a drowning man towards land and safety, the survival is not for itself, but for a different creature with which the porpoise feels an intuitive spiritual affinity.

Nests, hives, webs, dams, etc., constructions of birds, ants, bees, spiders, and beavers, are not planned goals, but are built according to species perpetuation and survival instincts. Man builds not only for survival and species perpetuation, but also to further his knowledge.

Chimpanzees and porpoises are highly intelligent – the ancient Greeks regarded porpoises as men of the sea – but they are races on their own. They are not men spirits starting or regressing in animal and fishes' bodies, but undoubtedly they will advance within their own sphere (to what limit?). Although there is a positive case for reincarnation, or metempsychosis, certain Eastern concepts of spiritual interchange between Man and animal bodies, does not seem possible. Man is (basically) too noble a spirit, and too far advanced.

Professor Theodosius Dobzhansky notes in his book *The Biological Basis of Human Freedom*, that:

> 'The chimpanzee is much superior to other non-human primates in memory, imagination and learning ability. Nevertheless, there is a vast gulf between the intellectual capacity of chimpanzees and of man. Symbolic responses can be learned by chimpanzees only with considerable difficulty, and their frequency fails to increase with age.'

It is not only a physical brain less in powers of cognizance, etc., than Man, in chimpanzees, but the spiritual being too, is a factor retarding chimpanzees from further intellectual development. Although Man has not yet learned to utilize more than one-billionth of his potential maximum brain-load capacity during a 70–80 year life-span, indications are that Man, as a more exalted spiritual being, may (possibly) be able to grow additional neurons – grey brain nerve cells – to store increased data, solely by application to study and learning, with his far higher spiritual aspirations for future progress. Experience and age does not increase the intellectual capacity of a chimpanzee beyond a certain point!

Notable intelligence is evident even among creatures with exceedingly tiny brains: – the harvesting grain storing ants (*Aphaenogaster*), the leaf-cutting ants (*Atta*) who cultivate fungus on stored leaves, and the dairying methods of ants who keep 'cows' (aphides), but *only* Man, of all entities, can *continue* to learn and accumulate knowledge and wisdom, even long after 100 years of age, and up to the point of physical demise. This erudition is *not* uselessly dissipated.

Specialized observations can be made in regard to lethal bacteria – microbes – viruses. Are these microscopic life-forms (of which the largest virus measures only 1/1,000,000th of a millimetre) *inherently evil – or, is germ warfare nothing new?* Did evil spacemen develop Anthrax, Asiatic Cholera, Bubonic Plague, Kala-azar, Malaria, Smallpox, Syphilis, Tetanus, Typhoid, Yellow Fever and other diseases in laboratories on other planets, for use in germ warfare to annihilate planetary populations, so enabling an easy victory in claiming new (planet) territories?

Germ warfare methods developed on Earth would probably render the soil so deadly that Man could not survive in these areas for hundreds of years, but some (presample) Cholera, Plague, and others, (may) have shorter life-spans once all living victims had been claimed, and so enable a more rapid 'takeover' of the (our?) planet. Or what remains of these diseases today may merely be atrophied remnants of ancient germ warfare once so deadly that entire planetary populations rapidly succumbed! It may be noted that virulent syphilis was even more lethal in the early 16th century!

Perhaps a clue can be found in the Holy Bible. What seems to mean either biological, or chemical, warfare or attack, or radiation poisoning, is described in *Revelations 11: 6 and 16: 1, 2, 3.*

11: 6 states:

> 'These (angels) have the power ... to smite the earth with all plagues, as often as they will.'

16: 1, 2, 3 state:

> 'Go your ways, and pour out the vials of the wrath of God upon the earth. And the first (angel) went, and poured out his vial upon the earth; and there fell a noisome and grievous sore upon the men (VD?) ... And the second angel poured out his vial upon the sea ... and every living soul died in the sea.'

Friendly angels these, who carry out orders under God's 'wrath'! If angels were space beings, it is obvious that not all were benign!

Were certain plant diseases also artificially developed on other planets? Remember that U.S. scientists produced defoliants, etc., for use in Vietnam!

The two subjects of philosophy and metaphysics could be expanded until completion of this book, but having reflected on different but related concepts and fashioned a substructure, it is now essential to return to the dissertation of physical/material discoveries and historical proof of early extra-terrestrial visitations to Earth.

Part 2, The Migrants

WAS MODERN EARTHMAN BROUGHT HERE IN SPACESHIPS, AND WERE THE 'APE-MEN' ARTIFICIALLY CREATED BIOLOGICAL ROBOTS?

In association with some of the theories outlined in this work, and continuing on from the interrelated subjects of philosophy/metaphysics, a sensational archaeological find was made in 1925 by a Dr. N. Grigorovich, at Odintsovo near Moscow.

Grigorovich discovered a yellow-brown 'fossil' of what unmistakably appeared to be a *human brain* embedded in carboniferous limestone sediment. This belongs to a period of pre-history so many millions of years ago (260–340 million years), that orthodox thought cannot possibly conceive of any human being ever having lived on Earth in that period.

Over the years since 1925, valiant and repeated attempts have been made by geologists to explain away the find in terms of orthodox geological thought, but even after all these years the 'stone brain' *still* remains an insoluble enigma; for this amazing fossil has, in its remarkably striking resemblance to a human brain, a longitudinal dividing fissure between the two hemispheres, more than 15 convolutions, plus the cerebellum and vermis cerebelli!

If this fossil really was once a human brain – and indications are that it was – to whom did it belong? A spaceman visiting Earth many, many millions of years ago?

Apart from a slight deformation caused by extreme terrestrial pressures over millions of years, the fossil is identical in appearance with the brain of a modern Earthman! If the Odintsovo fossil was once the brain of a visiting, or dwelling, spaceman, and it is the same in appearance as the brain of a modern Earthman, then the Theory of Evolution is obviously in extreme doubt. Did we come from 'out there'? Have we dwelt in our present bodily form, on other planets, for hundreds or thousands of millions of years? Were the ancestors of modern Earthman – Homo sapiens or 'wise man' – placed on Earth (more recently) by spacemen?

Professor Paul Santorini, Fellow of the New York Academy of Science, pioneer of radar, and close colleague of Einstein, believes that the three chief factors in the world blanket of UFO secrecy, are:

FEAR OF PUBLIC PANIC, BREACHES OF NATIONAL SECURITY, AND UPHEAVAL OF THE CHURCHES' ESTABLISHED DOCTRINE OF CREATION!

An important factor in the question of Earthman having an extra-terrestrial origin, is the extraordinary diversity of languages spoken on Earth. Could this indicate that many Earth peoples came from different planets, each with a different single language?

Anthropologist Carelton S. Coon, notes in his book *The Origin of Races* that of the thousands of languages spoken on Earth (more than 2,800), literally hundreds are 'unrelated to each other' (tonal, non-tonal, clicks).

Many archaic languages were actually more complex, involved, and expressive in antiquity than are those same languages today! Instead of starting simplified then developing through history, the opposite took place, i.e. they *began advanced* then simplified through the millennia! Anglo-Saxon, German, Scandinavian, and Chinese, are notable examples of these languages. Surely this could indicate arrivals on Earth of advanced language-speaking peoples whose speech degenerated through the aeons? Or perhaps the language degeneration was due to self-destruction of these persons' once superior civilizations after reaching Earth, then the surviving populace taking up the threads of the old language?

China, which according to legend was civilized 1¼ million years ago, is proven to have once spoken a complicated polysyllabic language which is now of a simplified monosyllabic structure!

'Life on Earth may have started when spacemen landed here billions of years ago.'

(Professor Thomas Gold, Cornell University.)

And the twentieth century:

'Something unknown to our understanding is visiting this Earth.'

(Dr. Mitrovan Zverev, leading USSR scientist.)

Certain fossil and other evidence indicates that modern Man 'suddenly' appeared on Earth, and *did not* 'evolve' from primitive hominoids, or other hominoids considered to be degenerate evolutionary branches in the so-called 'chain' leading to modern Man. This of course also casts doubts on the relationship of 'killer apes', etc., to modern Man.

Equally as curious, is that the fossil record proves that at least one of the hominoids – Neanderthals – also appeared suddenly (some 150,000 years ago) and disappeared more recently, while becoming progressively *more primitive* towards the end of their existence – the *exact opposite* of evolution!

A 200,000-year-old middle-Pleistocene skull of a previously unknown primitive hominoid, unearthed in July 1971 near Perpignan in the South of France, was immediately hailed as a 'key find', but was it? The skull is purported to 'fill the gap', as have earlier minimal discoveries, between Java and Neanderthal Man; but if the Pyrenean Man really was the preceding 'link' before Neanderthal Man then of course very abundant – not minimal – skeletal evidence of Pyrenean Man should exist, as it does for Neanderthal Man, but such abundant evidence does not exist. The also minimal Java Man skeletal remains, some hundreds of thousands of years older, though similar to other hominoids, were confined to Java, an island in the Indian Ocean isolated by sea and nowhere near France.

A weakness of the evolutionary theory is the insistence in attempting to link unrelated and diverse discoveries in a cohesive and unbroken chain to support this outmoded, naïve and dogmatic concept.

In his book *Early Man*, anthropologist F. Clark Howell makes this pointed and startling observation about Neanderthal Man skeletal finds:

> 'Neanderthal man ... abruptly disappeared. The evolutionary tendencies that he exhibited during this period are *extremely puzzling*. For he seems to have become *more "primitive", not less so*. The last fossils we have from Western Europe are even *squatter, bulkier, and more beetle-browed than their predecessors*.
>
> '... He was noticeably different from modern man and *became more so* as time went on ... In addition to stopping suddenly, the classic Neanderthaler is replaced with equal *abruptness* by *people like ourselves*. There is *no* blending, *no* gradual shading from one type to the other. It is as if modern man *came storming in* and dispossessed the Neanderthals – perhaps even killing them.'
>
> (my italics)

Why did the Neanderthals degenerate? Why did 'modern men' abruptly replace the Neanderthals? Could any degenerated Neanderthals – perhaps still further degenerated – have still survived? We will look at this possibility in a later chapter.

One site where the first evidence of later, i.e. 45 and 60,000-year-old, and more primitive, Neanderthal skeletal remains were found, was in the Shanidar cave in Iraqi Kurdestan's Zagros mountains, while Neanderthal skeletal remains from an earlier period of pre-history, discovered on Mt. Carmel in Palestine for example, possessed much more advanced characteristics (for a Neanderthal). Could there be any relationship between primitive hominoids and an enigmatic statement in the *Popol Vuh*, the sacred codex of the Quiché-Maya peoples, when it mentions 'several attempts' by the 'gods' to make Man? Were the 'gods' spacemen scientists, and the hominoids, the results of 'attempts' to make Man?

The next abrupt appearance – of major proportions – was the 'Modern' Cro-Magnon (Cro-mañ-yon) or Palaeolithic Man, *also* with no proven ancestral 'link', who appeared about 35–60,000 years ago, possibly in France (?) while remains have been found throughout Europe, the Middle East, and even in Russia. Although for a period Cro-Magnon Man undoubtedly co-existed with degenerating Neanderthals, there is absolutely no 'link' whatsoever between the degenerated Neanderthals and the abrupt appearance of modern Cro-Magnon men; this fact cannot be explained by evolutionists.

Cro-Magnon Man was a superior physical specimen with a larger brain capacity than modern present-day Homo sapiens. Cro-Magnon Man, with his fine features and long head, also possessed a high degree of intelligence as is now belatedly being discovered. According to Alexander Marshack, research associate at Harvard University's Peabody Museum, the Cro-Magnon 'Baton of Montgaudier', 370 mm long, of reindeer antler, appeared to have been engraved '. . . with the aid of a jeweller's magnifying glass'. Researcher Marshack also established that, 34,000 years ago, Cro-Magnon Man was keeping track of the lunar cycle by abstract symbols! The surprising sophisticated Cro-Magnon art, up to 35,000 years old, painted in strange, still-fresh, non-fading pigments, shows this race to have been highly intelligent.

In 1970, Soviet archaeologists made some very interesting discoveries in Upper Palaeolithic (Cro-Magnon) excavation sites, alongside the River Sungir, near Vladimir, north of Moscow.

The Sungir expedition, organized by the Academy of Sciences' Institute of Archaeology, and led by Professor Otto Bader, discovered in this 27,000-year-old Ice Age community, beads, bracelets, and rings carved from mammoth tusks (with metal implements?), a tiny, exquisitely-carved ivory horse, shirts, and trousers made from undressed skin, bark boots, caps decorated with bone beads and Arctic fox teeth, and a delicate bone needle, as slender as, and the same length as, a modern steel needle!

Several other discoveries, including deliberately preserved skeletons of two boys, astounded the experts, but the most startling discovery was the weapons – 17 javelins and a spear – all 2 m 40 cm long and quite straight, but carved from *curved* mammoth tusks!

How, the experts wondered, could curved mammoth tusks be straightened? There is no explanation except for the obvious proof that these ancient people seemed to be considerably more intelligent than anyone had previously suspected!

Professor Bader remarked that these discoveries had 'thoroughly shaken up our ideas of the Upper Palaeolithic'.

The Auckland Institute and Museum is in possession of a Cro-Magnon skull from France which bears unmistakable evidence of *skull/brain surgery* undertaken on a Cro-Magnon in antiquity! An opening about 40 mm in diameter has been deliberately cut through the frontal bone almost over the right eye, seemingly with a trepan! In our surgical procedures a trepan is a cylindrical saw, used to remove a circular piece of bone from the skull, for one reason to relieve pressure on the brain, but according to prehistorians the discovery of metal was unknown during the supposedly Stone Age era of the Cro-Magnons! Primitive bone or stone tools probably could not cut such a precision opening, and might even kill the patient. From where did the ancients obtain metal surgical instruments? Were the Cro-Magnons space settlers, Muan (from Mu), Atlantean migrants, or descendants of any of these beings?

Surviving Cro-Magnon traces can apparently be seen in characteristics in Nordic, Mediterranean, and other races, but where *every* other modern race came from, evolutionists cannot satisfactorily explain; i.e. from whence did the *pure blood* Chinese come? The Peking Man (Sinanthropus), a pre-Neanderthal hominoid who roamed the plains of China some 360–600,000 years ago, was so physically primitive that it stretches the limits of credulity to accept the idea that this early creature was in any way related to the highly civilized and refined Chinese (or any other race), who, according to their own legendary accounts, possessed an advanced civilization which would have been at least 650–890,000 years before the Peking Man hominoid appeared on the scene!

Can the archaeologists really say with certainty that no evidence exists to indicate an advanced civilization in China one and one-quarter million years ago?

The Dawn of Magic by Louis Pauwels and Jacques Bergier, states:

> 'Systematic archaeological exploration has been going on for little more than a century ... No serious exploration has been carried out in South Russia, China, or in Central and South Africa. *Vast areas* still preserve the secrets of their past.'
>
> (my italics)

Not all Chinese today believe in evolution for themselves, for a Chinese term derogatory to the white man, states:

'White men look just like the apes from whom they are descended.' (From *The Many Worlds of Man*)

Of course it was an Englishman who first conceived of the evolutionary theory – other, more ancient and older civilized peoples cannot accept this concept for themselves.

But even other peoples aside from the Chinese must have had a sudden origin, for an extraordinary discovery was reported by *Life* in 1951, which stated that *100,000-year-old* skeletal bones of true modern Man – Homo sapiens – were discovered on the shores of the Caspian Sea! *Life*, which described these bones as 'amazingly different' from Neanderthals, and other hominoids of antiquity, further remarked that the discovered bones of Homo sapiens were actually slightly older than the hominoids of that area!

S. M. Cole, Fellow of the Geographical Society, writing in the journal *Discovery*, November 1951, said:

> 'The discovery of a jaw of Homo sapiens type in early glacial deposits at Kanam on the shores of the Kavirondo Gulf in Kenya, however, does make it appear at least possible that *modern man* existed at an early date, though it needs to be stated that this evidence is still not accepted by some experts.'
>
> (my italics)

The period S. M. Cole referred to was at least half a million years ago!

An even more thought-provoking discovery was reported by biologist Professor Frank Lewis Marsh, who remarked in his book *Evolution or Special Creation?* (a pertinent title) that:

> 'Another example of tampering with the evidence was furnished by Dubois *who admitted*, many years after his sensational report of finding the remains of Java Man, . . . that he had found *at the same time in the same deposits, bones that were unquestionably those of modern humans.*'
>
> (my italics)

Java Man – *Pithecanthropus erectus* – or erect Ape-Man, lived *at least 500,000* to as far back as *700,000* to possibly *1,000,000* years ago. Modern Man was apparently co-existing with Java Man.

Professor Eugène Dubois (1858–1941), a Dutch Army surgeon, made these little-known but vital discoveries in Java in 1891–92, and thereafter was said to have maintained a guarded secrecy.

Biology Professor A. M. Winchester states in his book *Biology and Its Relation to Mankind*, that:

> 'There was a time when it was thought that perhaps modern man was a direct descendant of the Java man, the Rhodesian man and the Neanderthal man. As the evidence has accumulated, however, it appears that this is not possible, because some ancient remains of *true man* have been found which were *contemporary* with the remains of some of these other forms.'
>
> (my italics)

There is in existence the cranium of a very primitive Rhodesian man hominoid about 40–50,000 years old, with a mystery round hole about 12 mm in diameter in the side of the head. The opening remarkably resembles a hole seared-through with a LASER* beam! The rim of the hole is unusually smooth, which (could) indicate slight melting, then cooling!

*Light Amplification by Stimulated Emission of Radiation.

If modern Earthman co-existed with early hominoids as far back as 500,000–700,000 to 1,000,000 years ago (at least?) this obviously casts great doubt on the Theory of Evolution, as even by the most recent estimates (late 1969) modern Earthman is considered to have appeared in comparatively recent times, merely a fraction of a 1,000,000-year time span! These discoveries outlined here *do not* conform with the Theory of Evolution but they *cannot be denied.* Evolutionists may mismatch such bone finds.

Were the discovered prehistoric skeletal bones of modern Man, those of Earthmen, or spacemen? From whom we must be descended, from somewhere along the line; the bones are the same as ours! The bones discovered seem to be those of Earthborn space being settlers, as oriental legends relate that China and Japan were already settled and civilized further back than 1,000,000 years, but significantly, within the time recorded in the Hindu chronology for the arrival of souls now dwelling on Earth; i.e. over 3,888,000 years ago.

More than 3,888,000 years ago, but less than 4,320,000 years ago, the first spaceship load of Earth settlers was apparently brought here in a deliberately planned resettlement scheme; probably (if this was so), several migration waves followed the first. Evidence dating back to scores of millions of years, i.e. a 'tiled pavement' 12–26,000,000 years old, at Plateau Valley, Colorado, indicates spacemen dwelling on Earth in those far-off days, but no evidence exists for a deliberate large-scale settler migration to Earth beyond the dates mentioned above.

How do the hominoids relate with all this? Were early and later hominoids *biological robots*? That is, directed work slaves, capable of reproduction, but with limited self-volition, and intellectual comprehension aside from self-survival capabilities? Were the hominoids less-advanced souls trapped in bodies developed, manufactured, or created in laboratories on other planets by spacemen, then dumped on Earth as slaves to serve modern Man. Or did the hominoids have another, or additional, purpose?

It could be assumed that these early man-like beings were placed on Earth by spacemen to see how they survived in a terrestrial atmosphere/gravitation, etc., preceding the arrival of modern Man. This possibility cannot be ruled out, but indications seem otherwise, for discoveries already mentioned indicate, by the fossil record, that modern Earthman preceded Kanjera, Neanderthal, Swanscombe, Peking, and Chellean Man!

Perhaps the hominoids were created in terrestrial laboratories on Lemuria, Mu, Atlantis, China, and other sites?

I absolutely, unequivocally and without reservation, forward my belief from research, that modern Man – or nascent humanity – was in no way whatsoever related to primitive hominoids or 'killer apes', etc.; but within my theories this does not preclude the possibility that the hominoids were related to the apes! This may seem an ambiguous statement in view of the anti-evolution theories advanced in this work, but the pronounced Simian characteristics of the hominoids were pertinent in the possibility of ape–hominoid relationship.

The Theory of Evolution is based on the premise of the culmination of small favourable gene mutations, or 'natural selection' of favourable gene mutations, but the Nobel-Prize-winning geneticist Prof. H. J. Muller remarked on 'Radiation and Human Mutation' (*Scientific American*, November 1955) that:

'In more than 99% of cases, the mutation of a gene produces some kind of harmful effect, some disturbance of function.'

Also, natural gene mutations have been proved to be so infrequent, that an animal or human gene and its hereditary characteristics, have been calculated to remain stable for millions of years. And of course we have only to remember the surviving species of plant and animal life, basically identical to their fossil ancestors up to 570,000,000 years old.

So, these three basic facts emerge: the chances of

natural gene mutation are slim, but possible; gene mutation, whether natural or otherwise is, by observation, more than 99% aberrant retrograde or regressive deterioration, while gene mutation is mostly caused by one or both of two external agents, i.e. radiation and chemicals. This is proved in Man by radiation from the A-bombs dropped in Japan, plus the drug thalidomide.

Did spacemen, Lemurians, Muans, Atlanteans, Chinese, and possibly others, deliberately alter the basic hereditary gene DNA 'blueprint' structure in the ova and sperm of apes, by radiation and/or chemicals and/or pre-programmed factors, to create the hominoids? But if the hominoids were partially or temporarily superior to the apes, then how could degenerative genetic mutation explain this? The apes, but no early hominoids, survived, and the Neanderthals exhibited every sign of progressive genetic deterioration; while if the very strange creatures sighted today in the Caucasus and North America are degenerated Neanderthals, then they are at present indicating signs of reverting to apes!*

Natural, terrestrial or cosmic radiation could be considered as a possibility for genetic deterioration in the apes, hominoids, and of course the Neanderthals, but the improbability of this concept is explained in Chapter 5. The only other consideration for unnaturally high radiation, is nuclear warfare on Earth; but indications are that the hominoids were *deliberately* created by beings of high scientific intellect.

Let me explain further. In June 1970, the Indian Nobel Prizewinner, H. Gohbind Khorana, announced that he had achieved the first total synthesis of a gene, from atoms of simple chemical building blocks. *No* natural gene was required as a model in the reaction mixture. This, of course, is an important factor in the eventual possibility of creation of life-form bodies (or alteration of existing life-form bodies?). A gene is the factor determining hereditary characteristics, which is transmitted by each parent to offspring.

*Although rare, it is not unknown for a mutated gene to revert to its original 'blueprint'.

In November 1970, Dr. James Danielli, the leader of a three-man team of scientists from King's College, London, announced that in New York* they had successfully produced single-cell jelly-like amoebas artificially! From groups of three amoebas they took the nuclei, membranes, the cytoplasms, and synthesized them into living cells able to reproduce!

Dr. Danielli, who has been researching in this field since 1945, believes that within 10–20 years it could be possible to artificially produce new mammals, including people. The British biologist also forecast the artificial synthesis of new animals, plants and micro-organisms.

The amoeba research was undertaken by Dr. Danielli as part of a project of NASA to create biological robots able to live in possible harsh planetary atmospheres, but hasn't all this been done a long time ago?

As already outlined in this chapter, indications are that life-form bodies for plant, animal and human disease bacteria – microbes – viruses – may have been artificially created in outer space laboratories by possibly evil spacemen. If this is possible, then it must also be considered that early and later hominoids could have been biological robots, artificially created by scientists of pre-history.

Excavations in 1965 at the Molodovo site in the Ukraine, proved that Neanderthals had once built structures at that site, but why did they 'suddenly' appear without an ancestral 'link', then become progressively more primitive, before apparently 'suddenly' disappearing about 35,000 years ago? Did they, in fact, *not* become extinct, but in reality did they *escape* from the Cro-Magnons, and/or their own masters (whom, and for *what* reason?), and go into hiding in the Caucasus, Himālayas, Pamirs, Tibet, Chinese Turkestan, Central Asian deserts, Mongolia, and across a once-existing 'land-bridge' between the Chukotski Peninsula and Alaska into North America? Were they degenerated *still further* by progressive, degenerative genetic mutation, and survived to this day?

No prehistorian or evolutionist can now really

*State University, Buffalo.

seriously believe – even with rationalizations – that the Neanderthals were a direct 'link' in the 'chain' leading to modern Man, but everything exists for a purpose. Were the Neanderthals biological slaves for modern Man, or were the Neanderthals, and other hominoids, merely the results of scientific attempts to artificially create Man?

Were the Neanderthals created in Atlantean laboratories? Curiously, although the Neanderthals cannot possibly compare with the later Cro-Magnons, as advanced beings, France was the richest fossil graveyard of both; in this context, the Cro-Magnons are mentioned later. This seems to indicate likely entry from North-West Atlantis by placement or migration over a possible 'land-bridge'. For those who do not believe in the past existence of the Atlantean continent, Chapter 3 describes some extraordinary late 1960s discoveries in this connection.

Alternatively, did the Neanderthals *escape* from Atlantis, over a land-bridge, by entering France? Whatever the reason, the principal Neanderthal habitation sites do indicate a spreading-out from France over thousands of years: i.e., the approximate locations were, in France 15, Germany 7, Spain 4, Ukraine 4, Italy 2, and farthest from France, one site each located in the Crimea, Ethiopia, Libya, Spanish Sahara, Morocco, and other sites. Although Morocco is close to Spain, it is difficult to imagine the Neanderthals swimming the Straits of Gibraltar, or constructing boats. Excavation sites indicate slow land migration (or air-lifts?) from a central point.

Also notable in France are the megalithic 'menhirs' or 'standing stones' (raised by anti-gravity?) at Carnac, Brittany; mute testimony to an unknown, advanced (Atlantean?) civilization.

It seems extremely unlikely that low-intelligence Neanderthal Man 'evolved' on Atlantis or any other site, for in fact, the Neanderthals were polymorphic. In Africa, the considerably more primitive beast-like Rhodesian Man who lived (about) the same time, and who could not make implements, was also classified as Neanderthaloid, which indicates at least two bodily models.

Although (it seems) the most likely probability is that the Neanderthals were artificially created from apes on Atlantis, if they were brought from outer space, the degeneration might be explained by terrestrial factors – atmosphere, gravitation, etc. – unsuitable for creatures designed on other planets. But the Neanderthals could not have been degenerated spacemen stranded here, for the Neanderthals were not too advanced to start with – even with a brain capacity averaging 150 cc larger than our own – and could hardly be compared with a vastly superior spaceman, or modern Earthman.

A new concept is the theory that the Neanderthals suffered from (and were degenerated by?) rickets, induced by vitamin D deficiency – through dietary and/or sunlight insufficiency. But in the 22nd January, 1971, issue of *Nature*, Drs. Ernest Mayr of Harvard University and Bernard Campbell of Cambridge forwarded their belief that vitamin D shortage in Neanderthal Man was *unlikely*, due to the *geographically widespread* Neanderthal habitation sites, combined with the low latitude and *warm* climates. A different theory that osteoarthritis ravaged the Neanderthals, is also illogical as a possibility for *collective* racial ill-health of a widespread physical type. And of course, arthritis could not explain 'squatter, bulkier and more beetle-browed' Neanderthals! The two researchers Mayr and Campbell noted that Cro-Magnon Man who appeared at the peak of a cold stage, and who, incidentally, co-existed with Neanderthals, particularly in France, passed through several subsequent cold phases but did not exhibit any deterioration, or vitamin D deficiency symptoms due to lack of sunlight and/or diet.

If vitamin D was lacking in the diet of Neanderthal Man, which is improbable, then he should have been able to synthesize it by the same process that we can, i.e. our apparent built-in (by whom?) safety factor – vitamin D2 produced by the action of the Sun's ultra-violet rays on the oil glands of the skin, which then secrete the wax-like pro-vitamin irradiated 7-dehydrocholesterol, before final conversion into vitamin D. Only five minutes' daily exposure to midsummer sun, or just under three hours to weak midwinter sun, is sufficient to produce ample vitamin D in a test rat. So if the Neanderthals lacked this synthesizing ability – inherent in humans, animals and birds – then they must have been a very strange (artificial?) creation! Still another point indicating artificial creation of the Neanderthals, though, significantly, if rickets was the cause of the Neanderthals' degeneration, then they would have died out – become self-extinct – *but did they?* Do *still further* degenerated Neanderthals *still* survive – the Caucasus 'almastis', California 'Bigfoot'? This subject, with its ramifications, is examined in Chapter 11.

Many eye-witnesses have sighted small, peculiar-shaped beings near landed UFOs, whose sole purpose it seems is to carry out routine plant- and soil-gathering operations and other simple procedures. These 'goblin'-type beings, often described with divers variations, i.e. 'yellowish-orange glowing eyes', without a nose, mouth or ears, and covered with fur, or with huge ears, large eyes side mounted on the cranium, etc., do not exhibit very advanced qualities when sighted by often frightened people. Some reports refer to attacks by these tiny dwarf-like beings, but other sighting reports describe tall, medium and small space beings who resemble Earthlings, and who do show advanced spiritual qualities.

Are the first mentioned goblin-type, plant-/soil-gathering beings artificially created biological robots?

Research into particular branches of study indicates that anthropoid apes and other non-human anthropoid primates are, and were, so uniquely organized in certain physical characteristics – viz. the central nervous system – that the possibility of any of them ever having been in a related line of modern Man's 'descent' seems quite remote.

If modern Man did not descend from non-human anthropoid primates, apes or primitive hominoids, *from whence* did he come?

> 'Contemporary Hopi, Zuni, and many other Indian tribes, as well as prehistoric Toltecs and

Aztecs, believe in the myth that they lived on three successive worlds before coming to this one.'

(excerpt from book: *Pumpkin Seed Point* by Frank Waters.)

Is it myth or truth?

New Discoveries in Babylonia about Genesis by P. .J Wiseman, states:

'No more surprising fact has been discovered by recent excavation than the suddenness with which civilization appeared in the world. This discovery is the very opposite to that anticipated . . .'

Was modern Earthman resettled here in antiquity – brought from distant worlds in spacecraft? The African Baganda tribe preserve their pristine legend of the Creation in which Man is believed to have descended to Earth from 'Heaven'!

Several prominent theologians and clergymen now believe that Adam and Eve were not two separate entities, but instead, represent the whole of humanity.

It seems illogical to assume that the three (known) basic racial types – Negroid, Mongoloid, and Caucasian – could have 'descended' from just two people of undetermined racial type!

Even what seems like evolutionary proof is only inherent variation, necessitated by adaptability to climate, surroundings, etc., to dietary influence, or to DNA 'blueprints' already (pre-programmed?). For example: when a butterfly egg changes to a caterpillar, chrysalis, and finally to a butterfly, Creature variation does not change it into a *different* species.

During research into Polynesian legends for this book it shortly became apparent that one of New Zealand's Maori legends, not previously researched in the UFO context, alluded to a subject possibly pre-eminently akin with my own supposition. After close analysis the legend's implications were that a percentage of modern Earthman's ancestors came from the star (or a planet of?) Whanui or Vega, 26 light years from Earth! (or from Vega's constellation Lyra?)

Also of great and singular interest regarding the origin

of Earthman, is the thirteenth-century mosaic dome in the atrium of St. Mark's, Venice, depicting the 'Creation'. Towards the centre of the dome are portrayed what could only be UFOs! Two disc-like flying objects are shown, complete with circular central sections resembling pilots' cabins, and surrounding edges. Rays or jets are emanating from six points around the inner edges of both discs; angels are standing alongside. Other, similar, discoid and spherical objects are pictured in this mosaic, including a sphere encrusted with stars settled on the ground with angels or beings gazing out through two circular windows. The Gothic designers of this exquisite inlaid work evidently associated the flying discs and spheres with angels! *Were the angels UFO pilots – was Earth originally peopled in this way?*

Many races of old cherish legends relating to their own origin; some firmly believe that their ancestors arrived on Earth from distant worlds. One example are the Ainus, or Ainos, an ancient aboriginal people, now of small numbers, who live in the north of Japan's Hokkaido island, and on Sakhalin island. To this very day the Ainu believe that their heavenly ancestors arrived in Shintas (UFOs?) from the sky, and that their first god, and/or possibly ancestor, Okikurumi-Kamui, landed at Haiopira, Yezo, in a Shinta. The Ainu believe in immortality of the soul, and in a Supreme Deity.

Eskimo legends recount that millennia ago their ancestors were resettled from Central Asia and other sites to the far North by gigantic spaceships. Curiously, among the treasures preserved in the Leningrad Museum of Anthropology and Ethnography, are several finely tooled and extremely unusual walrus-tusk carvings of strange 'winged objects' excavated in the Chukotka Eskimo burial grounds. The delicately chiselled carvings are dated in the first centuries A.D.; but what they represent (spaceships?) is still an unsolved mystery.

An allied puzzle is the mystery of the Kets, a people numbering only 1,200 who live in the lower reaches of Siberia's Yenisei. Their language, totally unlike their neighbours', is strangely related with languages spoken in the north Caucasus, in Spain by the Basques, and in North America by the Indians. Who transported the Kets from where?

Recently, a Soviet archaeologist discovered a Stone-Age petroglyph of a Cosmonaut near the city Navoi in Soviet Central Asia. *Tass* reported that the graphic work portrayed the entity carrying under his arm a space helmet with antennae on top, while on the back of the being was an 'object' (rocket belt, etc?) enabling him to fly. The archaeologist remains firmly convinced that this prehistoric drawing really was representing a space visitor of antiquity. Does this drawing portray one of the spacemen who transported the Eskimos to the North in spaceships? Julia Sorokina, a Soviet ethnologist, discovered a 'link' between the Chukchi and Eskimo with the Polynesians; their blood grouping is similar, and even more startling, some of their legends are almost identical.

Were the Eskimos *really* air-lifted from several areas of our globe to the far North in spaceships? Central Asia was one point of origin for the Eskimos and is also the site for some fascinating archaeological discoveries; probably the most important find in this area (Gobi or Mongolian desert) was made in 1959. An early explorer in the Gobi was Marco Polo (1256–1323), but even he did not discover what was found in the Gobi in 1959. In that year, a joint Soviet-Chinese palaeontological expedition to the Gobi, headed by Chow Ming Chen, DSc, discovered a petrified shoeprint embedded in sandstone; the shoeprint, dated at *many millions of years old*, has rows of indentations across the print, roughly seven to each row, and several raised ridges running the full length of the print. The indentations and ridges suggest climbing boots as part of a space suit. Chinese and Russian palaeontologists who have studied this shoeprint of great antiquity, have been unable to offer an alternative explanation to the spacemen hypothesis. Serious researchers believe the shoeprint was pressed into the moist Gobi sand by a space traveller who landed *millions* of years ago!

Part 3

SONS OF GOD, OR INTER-STAR MISSIONARIES?

Who were the strange mouthless beings who, in times long gone, arrived in the Prince Regent River area in Western Australia – the isolated, inaccessible Kimberley Ranges – and left rock paintings of themselves which could be any age up to 12,000 years or possibly older?

Aborigines definitely did not paint them! The aborigines believe that the mouthless beings were painted far back in the 'dreamtime' (Creation) or 'beginning of things' by the mystery race and are, therefore, regarded by the aborigines with great awe!

Sir George Grey (1812–98) discovered the cave paintings in 1838, and noticed that they portrayed these enigmatic beings with haloes, giving them a divine appearance (did haloes originally represent space helmets?). Many of the Kimberley Ranges painted beings are shown with footwear having climbing studs similar to the Gobi Desert shoeprint! Some of the portrayed entities, which are over 3 m tall, have unusual characters in an unknown unidentified language on their headdresses!

Men and women are delineated with white skins (pipeclay) proving that they were a white or fair-skinned people. Several colours were used in these pictures: mostly ochres, vegetable dyes, and charcoal, but aborigines *do not* use so many colours together in their pictographic works.

Grey discovered from the aborigines that in their legends vague references were recorded relating to a man who came to Earth from the Moon (or another planet?), executed the paintings, then returned to the sky.

Through the years many theories have been advanced to try to find an explanation for the pictures of the mouthless, fair-skinned beings, but a satisfactory solution has not been arrived at unless the space visitors hypothesis is seriously considered.

Every one of the inscrutable beings is depicted wearing robes – some are painted in vermilion. The haloes are also unusual; it was not until the sixth century that haloes were accepted by Christianity for esoteric religious adornments in works of art, paintings, statues, etc., although further back Buddha was often portrayed with an aura around his head. Were the stylized haloes and mouthless appearance of the Kimberley visitors due to heads covered by space helmets? Were they a race of super beings – Cosmonauts – from distant worlds?

Of unwonted interest in the Kimberleys is the Geike Gorge region which appears to have been an *ancient defensive position*! What could this mean?

Many areas of Australia preserve ancient rock formations; though eroded and worn by time, some definitely seem to have been carved by skilled beings, in the shapes of birds, animals, etc. One of the most engaging sites is Mt Rufus in Tasmania, where giant rock formations resemble a falcon, Egyptian Sphinx, and so on.

The 3.6-km-high Marcahuasi Plateau in Peru has *strikingly similar* eroded rock formations to those on Mt Rufus! In appearance they seem about the same age. On this plateau are divers representations of human faces, a lion or sphinx, the Egyptian hippopotamus god Thoueris, and other animals and figures. Were the petroglyphs and carvings in Australia and Peru, executed by the *same artisans* from distant worlds – and/or Egypt?

WAS THE GARDEN OF EDEN A SPACE BEING/EARTHLING CROSS-BREEDING EXPERIMENTAL LOCATION IN CENTRAL AUSTRALIA?

Who was the very tall being who wore a conical (space?) helmet, and who was portrayed by an ancient artisan in a rock petroglyph in Central Australia, discovered in 1962 in the same area with an ancient human face engraving depicted with strange goggles and headgear – a space helmet?

Why would Central Australia be such an important location? A most unusual observation is that many ancient peoples have artificially duplicated by construction what appears to be smaller copies of the massive Central Australian Ayers Rock! Four examples are at Ur, the old Chaldean city on the bank of the Euphrates, Delphi, in India, and in Peru.

What possible reason would these historic races have to artificially construct replicas of Ayers Rock in Central Australia? What is the significance? Many years ago the famous author Col. James Churchward advanced the results of his research that the Garden of Eden was located on the (now) lost continent of Mu in the Pacific, but significantly ancient peoples cherished beliefs that the 'paradise' site, which *did not sink*, was marked by a massive rock in the same area of the world.

Situated 222 km to the north-east of Ayers Rock is the lush, fertile Palm Valley Oasis. Was is artificially cultivated by spacemen? Was this the real Garden of Eden?

Strangely, the man-made duplication of (Ayers Rock?) in some of the ancient lands mentioned, exhibits identical characteristics to the giant Australian Ayers Rock; i.e. precipitously sloping on one side with a more gradual declivity on the other! Aztec chronicles describe the same *unbalanced* giant rock mass in 'paradise' called 'Culhuacán'. From Palm Valley, Ayers Rock does appear to slope more steeply on one side, with a more gentle slope on the other!

It seems almost certain that Australia was once part of a much larger Southern Continent in the Pacific – either Lemuria or Mu, (also called MóO, or Pan). In the *Divine Comedy* Dante recorded two rather startling mysteries: one was an early Greek legend referring to a massive southern continent which sank, but left the remaining 'paradise' site; the other was a precise description of the Southern Cross.* But this constellation *could not* be seen

*If, as is thought by some, the Southern Cross was seen in high northern latitudes 5,000 years ago, it would have been seen, named, and catalogued by the Akkadians and Chaldeans, but no records exist.

from the Northern Hemisphere! Well, did the ancient Greeks know of the Southern Cross? How, and for what reason?

Possibly *one* of the most curious aspects of the 'paradise' enigma are Aztec legends which refer to 'paradise' as possessing a palm grove watered by four converging rivers. *Palm Valley is irrigated by four convergent rivers!* A very strange, weather-worn, castle-like rock formation is situated in Palm Valley; a rock formation so similar to many others in certain areas of Australia, which to the person without preconceived orthodox beliefs, resembles the work of, possibly, *giant* (space?) beings!

> 'And there we saw the giants – and we were in our own sight as grasshoppers, and so we were in their sight.'
>
> (*Numbers 13:33.*)

> 'For only Og king of Bashan remained of the remnant of giants . . . his bedstead . . . nine cubits* was the length thereof.'
>
> (*Deuteronomy 3:11.*)

Probable confirmation for the existence of Lemuria, or Mu, was that in April 1967 artificially carved rock columns engraved with *unknown* symbols were discovered at a depth of 1,806 m, close to the coast of Peru, by scientists on board the *Anton Bruun.*

Dr. Robert J. Menzies, the ocean research director of Duke University Marine Laboratories, said:

> 'We did not find structures like these anywhere else ... I have never seen anything like this before.'

The sonic depth recorder on the *Anton Brunn* detected 'lumps' on the level sea bed, which appeared to be the ruins of an ancient city in the same area. It seems certain that an advanced civilization was living here when it was dry land. Lemuria or Mu?

Though skeletal fragments of extreme antiquity, of modern Earthman, have been discovered at such

*About 4 m 07 cm.

locations as the shores of the Caspian Sea in the USSR, the shores of the Kavirondo Gulf in Kenya, and in Java, *the richest deposits of the prehistoric bones of modern Man may never be found!* Lemuria, Mu, and Atlantis, the legendary continents long since submerged, were claimed, according to legend, to be the first sites where *modern*, not evolving, Man appeared on Earth. Man did not 'evolve' on Lemuria, Mu, Atlantis, and China; aside from the striking similarity of the Odintsovo fossil 'brain' to the brain of modern Earthman, legends from Africa, east and south-east Europe, Japan, the North-East Frontier of India and other lands, including material discoveries, all testify to the arrival of fully modern Man from space.

The sudden appearance of Cro-Magnon Man suggests a migration from Atlantis to southern England and France of modern Man who had, perhaps, been dwelling on Atlantis for several hundred thousands, or even millions, of years; having originally been brought to Earth in spaceships. The Cro-Magnons were an enigma; they appeared to lead simple lives, but new discoveries prove that they possessed a high degree of intelligence. Perhaps the Cro-Magnons were outcasts from Atlantis, or people forced to migrate due to circumstances, or people desirous of leading simple lives – beings with advanced intellect but without all the material benefits of an advanced technology? And did they possess Atlantean implements, or receive assistance from spacemen or visiting Atlanteans? The sub-miniature engravings on the 14,000-year-old Baton of Montgaudier would require a magnifying lens for the engraver to execute his work; cranial trepanning would necessitate anaesthetic and metal surgical instruments in order not to kill the patient. And how was curved ivory tusk straightened?

Interestingly, the Cro-Magnon cranial/skull shape was similar to that of ancient Egyptians. Who themselves were associated with Atlantis?

Returning again to the Pacific Ocean – Lemuria and Mu – a discovery possibly in this connection was that in 1964 part of the mandible and teeth of a 200,000,000-

year-old ichthyosaur, a sea reptile, was unearthed *inland* in New Zealand's Hokonui Hills, Southland. This find could indicate that New Zealand or part thereof was at one time submerged before rising from the sea, or that a gigantic tidal wave from the sinking of *portions* of a large Southern Continent – of which New Zealand may have been a part – washed across New Zealand depositing sea creatures inland!

Chapter 6 outlines a Maori legend referring to a submerged New Zealand rising from the sea in antiquity. Did something violent happen during this period? At approximately the same time that the ichthyosaur was deposited in the Hokonui Hills, sudden, but not total, mass extinction took place at the close of the Palaeozoic or Permian era some 230,000,000 years ago. (When dinosaurs appeared?)

What could cause major upheaval on Earth at this time – continental drift, cosmic factors – or a nuclear holocaust? Notably, the Odintsovo fossil brain is dated near this period.

During my own intensive research into vanished Pacific continents, 'Sons of God' legends, Ayers Rock, aborigine legends, and 'serpent' deities of antiquity, it was of considerable interest to me to note that aborigines regard Ayers Rock as a 'holy place' associated with 'Creation'!

Even more startling is a religious symbol cherished by the Dieri tribe of *Central Australia.* This symbol is a Latin Cross identical to that connected with Christianity – i.e. a vertical beam with a horizontal cross-beam one third down. In all probability the Dieri tribe have known this religious symbol long, long before it was linked with Christianity!

Aside from the cross being used in antiquity as an instrument of death and disgrace, the cross as a religious symbol was known long before Christianity; in Egypt it was the 'ankh' – 'crux ansata' – while the ancient Gauls, peoples in South America, and other lands, also knew this symbol. Why did the Dieri tribe own the sacred cross symbol in Central Australia long before Christianity?

Did the (Latin?) cross *originally* signify or symbolize cross-breeding between spacemen and Earthlings in the Garden of Eden (Palm Valley?) and other lands?

Babylonian, Hebrew, and Egyptian sources of old referred to the Garden of Eden site as being in the 'underworld'. Is it possible that the 'underworld' referred to by these ancient sources really meant *Australia*, which is referred to today by peoples in lands far distant as 'down-under'? For the enlightened ancients living in the Northern Hemisphere, the 'underworld' may indeed have meant the Southern Hemisphere!

The concepts outlined here would of course be difficult but not impossible to prove in whole or in part, and undoubtedly there must remain many unknown or undiscovered factors. My theories can only be presented as a hypothesis for consideration, but while being probably much closer to the truth than the unconvincing, untenable, orthodox Judaic/Christian approach to the Garden of Eden legend, the theories outlined are also probably more acceptable to the freer-thinking, science-orientated mind of twentieth-century Man.

Rigid views held by theologians also extend to the Garden of Eden legend. Scholarly theological debate relating to this subject does seem rather ludicrous when naïve dispute has taken place over whether the allegorical 'apple' was, in reality, an apricot or a quince! The real location, scope, and implication of the Garden of Eden is above the comprehension and capabilities of orthodox theology; it is a question for science.

Continuing ... old Sumerian, Babylonian, Egyptian, and Greek legends refer to 'serpent' deities who were believed to have once resided in the 'underworld'. The Garden of Eden now takes on an additional interest and significance, possibly of paramount importance; i.e. the Garden of Eden, and 'serpent' deities/beings, both of which once existed in the 'underworld'! If in fact pristine legends from Australia and the Pacific islands are researched, one finds innumerable references to serpent deities/beings, who were anciently associated with the 'Creation' enigma in the area!

This now brings us to the next stage, i.e. the 'Sons of God' (*Genesis 6:2, 4*) who, in a way, were linked with what appears to have been intentional interbreeding with Earthborn women, as evidently were the serpent beings. A certain legend from the British Solomon Islands infers deliberate interbreeding between the serpent deities and Earthlings. Were the serpent deities/beings and 'Sons of God' the same race?

Some legends refer to the Sons of God as very tall fair beings, supposedly the descendants of Adam and Eve, but the obvious Biblical inference is that the Sons of God (Hebrew 'Elohim', seven* in number) were beings not of this world! Mummified and skeletal remains of a very tall race have been discovered in the Australasian/Pacific region.

Were the Sons of God the very tall white-skinned visitors to the Kimberley Ranges? These seem undoubtedly to have been spacemen. A few of the Kimberley Ranges beings were painted as tall as 3 m 15 cm. The mystery characters resembling the letters GITIL on the headdress of the largest figure cannot be identified with any specific Earth people. Do these characters depict the written language of another planet?

The spiral serpent symbol found throughout the Pacific is associated, as with legends, with the Creation enigma, while the aborigines have a most interesting name for Ayers Rock; in addition to being associated with Creation, they call Ayers Rock 'Uluru' which means 'serpent of the rock'.

Among many ancient peoples, serpent beings were believed to be space beings of infinite wisdom and knowledge who came to assist and teach Earthman!

Was the Biblical 'serpent' in the Garden of Eden in reality many serpent space beings (or Sons of God?) in Central Australia and possibly other lands, who interbred with Earthborn women to produce a superior genetic strain among their offspring? Greek and Indian legends of old refer to a *special* race or races of space beings who married Earthlings to produce superior offspring!

*or more?

Very ancient optical lenses were excavated in Central Australia some years ago (see *The Dawn of Magic* by Louis Pauwels and Jacques Bergier, p. 119). This discovery definitely indicates the existence of an advanced civilization in the Central Australia area!

Were the Kimberley visitors serpent space beings? In the Kimberleys are Baobab trees of a species found nowhere else on Earth! Baobab trees are found in Africa, Madagascar, India, Ceylon, and other lands, but the species found in the Kimberleys is unique. The botanical name is *Adansonia gregorii.*

How the Baobab trees became established in the Kimberley area is a major botanical mystery; they appear so alien to the Australian bush that it has even been suggested that they may have come from another planet.

Did the Kimberley visitors bring the Baobab trees from another planet? The Tonga priests of Africa regard their species of Baobab as 'holy'!

Whether the tall, fair Kimberley visitors were serpent space beings – probably Sons of God – is uncertain, but from available evidence it appears that they were one and the same. Biblical chronology for certain events of vast antiquity has, of course, been proven highly inaccurate, hence the possibility of the serpent beings and the Sons of God being the same!

Aborigine legends indicate that the serpent beings were not above waging war. Very odd, this identical parallel is also mentioned in the Hindu legends of the serpent beings – Nāgas* – who came from one of seven worlds (or planets? in which solar system?), who were concerned and benign towards Earthlings, and who were said to resemble Earth people; yet at the same time, they were capable of annihilating the countryside by brilliant super-hot (atomic?) means. We must remember, too, that at this period of history Australia was supposedly iso-

*Were *all* serpent space beings fair-skinned? Although possessing no technological attributes, not of fair countenance, and of non-Aryan ethnic origin, was there any possible remote association with the Tibeto/Burmese people called Nāgas who live in East Assam between the middle Brahmaputra and Chindwin?

lated from the rest of the world, yet we find identical legendary parallels!

Apparently, from aborigine legends, the serpent beings waged many wars (with whom?) around Ayers Rock. The 'vertical gutters' and 'potholes' in Ayers Rock are supposed to testify to these (atomic/LASER?) wars, according to the legends, while it can be noted in this context that the Kimberley's Geike Gorge region resembles an archaic defensive position!

Indications from aborigine legends are that the serpent beings utilized quartz crystal to release both destructive and healing rays. These rays may have been ultra-violet-light radiation; excessive UV-light radiation is dangerous and harmful to life-forms, while under carefully controlled conditions its effect is beneficial.

PERHAPS THE SERPENT BEINGS LIVED NEARLY SIX LIGHT YEARS AWAY?

Why did the ancients associate a serpent with extra-terrestrials? Surely these space beings did not resemble snakes? By the end of this chapter, we will discover a, or the, likely reason why.

The ancient serpent symbol is to be seen in many parts of the world; undoubtedly, the most fascinating portrayal is a detail on an Egyptian 'magical' papyrus in the British Museum which depicts a serpent encompassed by a disc emitting rays. Very old Olmec cave paintings in Mexico, dated between A.D. 600–400, picture humans and serpents in similar colours to the Kimberley Ranges paintings of the mystery haloed beings.

The most usual form of the serpent symbol is a spiral representing a coiled snake; it has been discovered as petroglyphs and other pictorial representations, in Britain, Greece, Malta, and Egypt. In New Mexico as pottery designs; as ground drawings on the Nazca Plateau, Peru, and throughout the Pacific Islands, which may possibly include the finely-carved spirals in Maori art.

In Australia it is the 'Creator' 'Rainbow Snake' of the

aborigines; in Iran, the Azhi Dahâka of the Zoroastrian religion and in Africa the symbol is reversed by the Zulus and the Maravi. It is also to be seen at Tassili-n-ajjer in the Central Sahara, while the Semitic serpent is known to most people.

Teutonic and Scandinavian deities or Aesir, were serpent beings, while in India they were known as Nāgas, whose king was named Vāsuki. The Druids, who taught that other worlds were inhabited, regarded themselves as serpent people, while the sages of ancient Egypt and Babylonia believed themselves to be 'sons of the serpent god'.

Let us now look at a possible origin for the ancient serpent beings. During the Christian era, and even before, there was in existence a religious sect called 'Gnostics' who possessed mystical knowledge of 'divine origin'. One early branch of Gnostics named themselves 'Ophites'. This name was derived from Greek 'a serpent', or 'Naasenes', from Hebrew 'nachash', which also means 'serpent'. The particular religious veneration of the Ophites was channelled into worshipping the 'serpent', or erudition introduced into the world by the 'serpent' at the Garden of Eden site.

Ancient peoples referred to the serpent beings as possessing great wisdom and knowledge:

> 'Be ye therefore wise as serpents.'
>
> (From a speech by Christ to His twelve apostles. *Matthew 10:16*.)

New Ireland native legends say the serpent beings provided food, while legends from San Cristobal Island say the serpent beings were 'creators'. Or were they 'Sons of God' space beings who interbred with Earthborn women? The ancient Egyptians, Hindus, Mayas and Quichés also regarded the serpent beings as 'creators'.

Papuan legends relate that the 'Gainjin' (Sons of God, serpent beings?) *came from a world in space identical to Earth*, to which they returned after their work on Earth was finished. The Kimberley Ranges beings returned to the sky!

A certain Melanesian people say the 'Kasa Sona' (Sons of God?) had arrived on Earth long before the first appearance of their tribe in the area.

We still have not arrived at the reason why the ancients associated the serpent with space beings. A little thought and investigation will show that in this connection there exists an important constellation or star group which extends from Northern latitude 15° to Southern 10° approximately and is known as Serpentarius 'the serpent bearer' or Ophiuchus (from the Greek 'Ophites' – 'a serpent', or 'Ophidia' – 'snakes'). Discovery of this constellation is *not* recent but, interestingly, was *well known* to the ancients. In Alexandria, Ptolemy, or Claudius Ptolemaus, the great astronomer, geographer, and mathematician, catalogued Ophiuchus between A.D. 127–145, while this star group was mentioned even further back by Eudoxus in the fourth century B.C. But *why* was it associated with a serpent? Significantly, Ophiuchus is a constellation in proximity to the winding serpent-shaped Serpens or Serpent constellation – i.e. one part of Serpens (Cauda) is just south of Ophiuchus, the other (Caput) is to the west. That is, Ophiuchus is between the head (Caput) and the tail (Cauda).

A most fascinating star is situated in Ophiuchus, and is of particular importance as a point of origin for the serpent space beings.

Barnard's Star, magnitude 9.5, almost 6 light years from Earth, in Ophiuchus, was discovered by E. E. Barnard in 1916, who also saw unusual flying objects one year earlier. These were reported in *Proceedings, National Academy of Sciences*, 1915.

In 1963 Dr. Peter van de Kamp, director of Sproul Observatory at Swarthmore College Pennsylvania, confirmed that a planet first ascertained by himself in 1944 was indeed in orbit around Barnard's Star. More recently, Dr. van de Kamp calculated a *second* planet around this star. The two planets have masses of 0.8 and 1.1 of Jupiter's mass.

Further, the two celestial bodies have been designated B-1 and B-2 with periods of revolution around the primary of 12 years for the smaller, and 26 years for one revolution of the larger planet. It has been theorized that as many as ten planets (could) be in orbit around this star. Is this where the serpent space beings came (and come?) from? And what was the *real* meaning associated with the symbolic seven-headed serpents of ancient Cambodia, India and Mexico? Did these stylized representations *really* signify a star in Ophiuchus with six/seven planets in orbit? And/or did they symbolize the seven or more Elohim – Sons of God/serpent space beings?

SUMMARY

To summarize, my belief from research is as already outlined. The physical universe and physical Man, etc., came into existence in the material sense, by a means totally alien, unconsidered and probably incomprehensible to orthodox theology and science. Moreover, my tenet is that the first men and women on Earth were brought to this globe – particularly Lemuria, Mu, and Atlantis – as settlers, from possibly, or probably, many different planets in many different solar systems, at least 1,750,000 to as far back as about 4,000,000 years ago!

The old Slavic fairy tale called *How Humans Appeared on the Earth*, states:

> 'Man was created away from the Earth a long time ago. When his world was about to end, God, in order to perpetuate human-kind, ordered that the angels* take several human couples to the earth to propagate. The angels scattered the humans, and wherever couples landed people have been multiplying since. Perhaps when the world comes to an end, God will again take humans to some new place for further propagation.'
>
> (From the *Fairy Tales and Stories of Podolia*; compiled by Mikola Levchenko between 1850–60.)

*spacemen?

In the Nihongi or Chronicles of Japan for the year 667 B.C., it is proclaimed that the first Japanese 'Heavenly Ancestor' arrived on Earth more than 1,792,240 years ago. Space-being settlers, and in this context the ancestors of present-day Earthmen and -women, may have been placed in different geographic locations which would suit that particular racial type regarding climate, atmospheric variations or conditions, etc. – i.e. matching as closely as possible the conditions of the home planet.

Again, what of the possible significance of Barnard's Star in relation to the serpent in the Garden of Eden? Positive proof that intelligent beings live on a planet or planets* of Barnard's Star has not, at this present time, been established; but indications from venerable legends, in addition to recent discoveries of planets in orbit around this star, strongly tend towards this possibility – or probability. The British Interplanetary Society reported in 1970 the suggestion to place a powerful radio beacon on the Moon to announce the presence of Earthmen to other civilizations in space. The stars of particular interest to the society are: Barnard's; 70-Ophiuchi, a binary 16½ light years away in Ophiuchus; and the binary 61-Cygni which is known to have at least one planet in orbit.

Working on the theory that intelligent beings *do* come from a planet or planets of Barnard's Star and/or 70-Ophiuchi, and were the serpent beings in the Garden of Eden (Palm Valley), then they probably arrived in Central Australia within the last 12,000 years. Assuming that the first men and women were placed on Earth up to about 4,000,000 years ago – as recorded in the Hindu chronology – this would mean that people in the Pacific were living on Lemuria and Mu, and some crossed over a 'land-bridge' into Australia in comparatively recent times.

The (modern?) aborigine is known to have been in

*As the planets have only been postulated by mathematics and not seen, and as no atmospheric/planetary probes have been sent to these bodies, scientists/astronomers cannot at this stage definitely assert that no life exists there.

Australia some 12,000 years – migrating from where? Southern India? Mu? Or from Mu into India (the Dravidian Gonds?) and Australia?

The 12,000-year date corresponds with discoveries outlined by Col. James Churchward in *The Lost Continent of Mu* when he established that Mu was submerged 12,000 years ago. Churchward did not consider that Mu embraced Australia, but Aztec/Greek legends indicate that a Southern Continent did in fact embrace Australia, which could explain the appearance of the aborigines about the same time that a large portion of a Southern Continent sank. Were they transferred by UFOs?

In 1969, two complete, and five incomplete, mineralized and calcium-carbonate-encrusted skeletons – two females and five males – of a primitive hominoid at least 20,000 years old, and closely related to the Java Man, were excavated 193 km north of Melbourne in a sand dune skirting the Kow swamp. Were the hominoids Lemurian or Muan biological robots?

Mr. A. G. Thorne, anatomy lecturer at Sydney University, said:

> 'I feel the features in the Kow Swamp skeletons – as it is known – which are not seen in modern aborigines, could not have evolved into the modern aborigine.'

So it seems the aborigines really did come from elsewhere but morphologists have noted that the modern aborigine is the *merging of two or more physical types!* A Muan race with other Earth races – *and* serpent space beings? A taller, fairer, bloodstrain exists in Central Australian aborigines, while the Ooraminna Central Australian rock carvings 6–10,000 years old of figures 'wearing strange head ornaments', probably space helmets, actually pre-date aborigine culture in this area. So the theological date of circa 4,000 B.C. for the Garden of Eden may, in this case, apply to Central Australia.

Expanding these theories still further, the serpent space beings – probably 'Sons of God' – interbred with some Earthborn women but not all, in Central Australia's Palm Valley, and other parts of the world where Earthlings could be ameliorated by the infusion of superior genetic characteristics.

Even today, fairer-skinned natives on certain Pacific islands show the infusion since antiquity, of the blood-strain from a race of very tall, possibly Caucasian, white men – apparently taller than any Earthmen. The prehistoric Australian rock petroglyphs and paintings, if life-sized, prove these tall beings to have been from over 2 m 10 cm to 3 m 15 cm tall!

The hoary weather-worn artificially-shaped rock carvings in Tasmania – Egyptian Sphinx, Falcon, etc. – and petroglyphs at other Australian sites, evidently prove visitations from Sri Lanka (Ceylon), Egypt, and possibly from Assyria and Indonesia, which in itself is a remarkable discovery, but even this does not explain the very tall white race who were, it seems, the Kimberley Ranges' haloed, mouthless, probably space-helmeted, beings.

Returning for a moment to the possibility of ancient Egyptians in Tasmania – Mt. Rufus – consider, even using the shortest routes, the vast distances they would have to travel. Even in a direct line across the Sudan, Ethiopia, Somalia, the Indian Ocean, south-west Australia to Tasmania, or by overland and sea through Saudi Arabia, Persia, India, Sri Lanka, Sumatra, Java, Timor, into the Northern Territory of Australia, overland to Victoria, then across to Tasmania. By conventional ancient means of transport it would have been a formidable journey to reach Tasmania. Did the Egyptians travel by a faster method? UFOs?

The Egyptian magical papyrus already mentioned, now kept in the British Museum, was, according to legend, used as a 'guide' on a journey to the 'underworld' (Australia?). The Egyptians, of course, lived over the Equatorial line in the Northern Hemisphere. Also, it must be remembered, is the portrayal on this papyrus of the serpent god encompassed by a disc emitting rays (or jets?)

A most unusual human skull unearthed in New Zealand early this century, was later identified as ancient

Egyptian by Sir Grafton Elliot Smith, a leading Sydney anthropologist and anatomist.

That the serpent space beings were also 'Sons of God', is indicated by a statement made by the prominent Soviet physicist Matest Agrest, who believes that superior spacemen from any non-specific (and specific?) corner of the universe, could have been classified by the ancients as 'Sons of God'.

Evidence that interbreeding between the Sons of God and Earthborn women in different parts of the world produced genetically superior offspring, is in *Genesis 6:4*, where it states that the offspring of such a union were 'mighty men, which were of old, men of renown'.

There is no evidence that the very tall, fair space beings – 'Sons of God' – interbred with every race on Earth; the Chinese are a pure-blood race and have remained that way since antiquity.

I do not believe that it is possible at this stage to solve every detail in the origin of Man, Garden of Eden mystery, due to the enormous amount of research required, with only slender facts available. Facts which have become so interwoven with myth, mystery, misinterpretation, incorrect reporting, recording and evaluation, that it would take many years of research – particularly in the Pacific region – to attempt to solve every detail. But in this chapter, largely through available evidence in the Australasian/Pacific area, it is possible that fragments of the truth of what *really happened*, fit in closer to the mosaic of reality, than any other theory advanced relating to the origin of Man, Garden of Eden, etc.

This is my 'space-age' hypothesis for the origin and terrestrial distribution of mankind, and for the Garden of Eden legend.

End of hypothesis, and summary.

2. METAZOAN ANIMALS SUDDENLY APPEARED 570,000,000 YEARS AGO AND FLOWERING PLANTS 100,000,000 YEARS AGO

'Innumerable suns exist; innumerable earths revolve around these suns ... living beings inhabit these worlds.'

(*Giordano Bruno*, 1548–1600)

Not only Man, but animals and plants *also* it seems, appeared suddenly on Earth in antiquity. What appears to have been the definite emergence of a *pattern* for the introduction of life-forms on our planet (by whom?) gradually took form!

Going even further back, long before the appearance of modern Earthman, highly complex and advance life-forms – skeletal – 'suddenly' appeared on Earth in the Cambrian period starting 570,000,000 years ago. No pre-Cambrian fossil evidence exists to show an evolutionary transitional stage from earlier times; the life-forms – including flora – 'suddenly' appeared. Beneath the Cambrian strata are enormous empty thicknesses of sediment where no fossils of complex life-forms can be found.

Cambrian rocks are found mainly in Wales, Scotland, Ireland, Europe, North Asia, and North America, in strata or layers 903 m to more than 9 km deep.

Was the Cambrian period a selected time when cosmic beings introduced certain life-forms on our planet?

In *The Great Infra-Cambrian Ice Age* by W. Brian Harland and Martin J. S. Rudwick (*Scientific American*, August 1964), the authors state that:

> 'Both the sudden appearance and the remarkable composition of the animal life characteristic of Cambrian times are sometimes explained away or

overlooked by biologists. Yet recent palaeontological research has made the puzzle of this sudden proliferation of living organisms increasingly difficult for anyone to evade . . .

'These animals were neither primitive nor generalized in anatomy; they were complex organisms that clearly belonged to the various distinct phyla, or major groups of animals, now classified as metazoan. In fact, they are now known to include representatives of nearly every major phylum that possessed skeletal structures capable of fossilization; . . .

'Yet before the Lower Cambrian there is scarcely a trace of them. *The appearance of the Lower Cambrian fauna . . . can reasonably be called a "sudden" event.*

'One can no longer dismiss this event by assuming that all Pre-Cambrian rocks have been too greatly altered by time to allow the fossils ancestral to the Cambrian metazoans to be preserved . . . even if all the Pre-Cambrian ancestors of the Cambrian metazoans were similarly soft-bodied and therefore rarely preserved, far more abundant traces of their activities should have been found in the Pre-Cambrian strata than has proved to be the case. Neither can the general failure to find Pre-Cambrian animal fossils be charged to any lack of trying.'

(my italics)

This is the first – as proved by the fossil record – appearance of advanced life-forms on our planet. Understandably, even Darwin was mystified by the Cambrian Period.

From the time Earth first came into existence 4·5 to 11 billion years ago, as theorized by E. K. Gerling, a period some 570,000,000 years ago may have been conducive to certain life of the animal kingdom, after Earth had cooled – or warmed – sufficiently, and developed an atmosphere.

Did spacemen begin to populate our world with certain

species of fauna at this time? And what of the 'missing links' between birds and reptiles, reptiles and mammals, mammals and quadrupeds and fishes, and fishes and vertebrates and invertebrates, and so on down to protozoa? The Evolution doctrine does not supply the vital answers to these and other questions. Did a 'Noah's Ark' spaceship bring the above-mentioned creatures to our globe?

As described, 3.2-billion-year-old algae plant cells – evolutionarily unchanged to this day – were discovered in Africa. Was this *also* a deliberate instigation by extraterrestrials to pave the way for the later introduction of the metazoans, or multi-cellular animals, 2,630,000,000 years later? For example, by the process of photosynthesis the algae would convert large quantities of inorganic carbon into organic carbon, which would form part of the bodily chemical make-up of reproducing life-forms. The algae would also assist in oxygenating the sea and atmosphere, and would disseminate chlorophyll; all prerequisites for the introduction of certain life-forms 2.6 billion years later, while the algae would also be utilized by life-forms as food.

In more recent times, some of the sacred, and other, creatures of past civilizations are surrounded by a romantic aura relating to their first appearance. A Theosophical belief is that the honey bee was brought as a gift to Earthlings by Venusians. The tiny Chihuahua dog was revered and regarded as sacred by the Aztecs of old. The mouse, roe, snake, and swan were sacred in ancient Greece, and identified with the demi-god Apollo.

The non-barking Basenji dog has been traced back in lineage to ancient Egypt, while it can also be noted that the Egyptians of antiquity worshipped the Slughi greyhound, and their sacred cat, or 'Mau'. The 'Mau' was under the special protection of the goddess 'Bast', 'Bastet' or 'She of Bubastis' in the Nile Delta. The 'Mau' was worshipped as far back as 4,000 years ago, while portraits have been found dated in the vicinity of 1600 B.C. The modern descendant is the Abyssinian civet cat.

Another creature of interest in this context is the horse, which was associated with the Sun by the ancient nomad horsemen of South Russia, and other peoples.

Were the bee, chihuahua, horse, mouse, roe, snake swan, basenji, slughi, and mau, and others too numerous to mention, such as Ganésa the sacred elephant of the Hindus, all brought to Earth from other planets?

DID KUAN-YIN BRING RICE FROM THE PLEIADES?

One hundred million years ago, with abruptness, the 'age of flowers' began. Before that time, no angiosperms of flowering plants bloomed on Earth!

Candolle (1778–1841), the famous Swiss botanist, recorded that the origin of 32 cultivated plants is unknown, while there is a doubtful origin for another 40. Among the plants with an unknown origin are the Imperial Lily, Lilac, Tuberose, and certain roses. Did these flowering plants originate on Earth, or another planet?

The following examples are only a few, of sacred trees, plants, flowers, foods, etc. Why did so many ancient peoples worship them with such devotion and reverence? Is it possible that these sacred items of flora were indigenous not to Earth, but to another planet, or planets? Were they in fact, 'gifts from the gods'?

Brazil	Manioc, Mandioc, Manihot, Cassava or Yuca plant; sacred to the Urubus.
China	Rice: brought to Earth as a 'gift' by the goddess Kuan-yin. Chrysanthemum: Hoso, the Great Grandson of the Emperor Juikai, was said to have lived 800 years without any bodily deterioration, because he drank the 'dew' of this 'holy flower.

Egypt	The date palm, garlic, onion, leek, the hyacinth and the sycamore tree were all venerated or dedicated to the gods by the ancient Egyptians.
England	Holly, mistletoe, and the oak tree were revered by the Druids. They regarded mistletoe as a 'gift from heaven'; it was considered more sacred than anything in the world.
Greece	The bay tree and the date palm were sacred to the deity Apollo. The lime tree, poppy, and the rose were sacred to the goddess Aphrodite. The pomegranate was sacred to the goddess Hera.
India	The pipal, peepul, bo or bodhi tree, the *Ficus Rumphii*, and the banyan tree: the pipal and *Ficus Rumphii* fig trees are sacred to the Brahmins and Buddhists, while the banyan is sacred to the Hindus. Still other items were the 'forget-me-not' flower believed by the Brahmins to have come from paradise, and the soma plants. The Rig-Veda-Samhitâ Vedic hymns record that the god 'Soma' was brought down to Earth by a falcon – or UFO? The two varieties of soma plants, *Asclepias acida* and *Periploca aphylla*, are sacred in India.

Indonesia	Rice: According to an Indonesian legend, rice was brought to Earth from the Pleiades (by Kuan-yin?).
Japan	The chrysanthemum, plum, and cherry were venerated by the Japanese of old.
Mexico	The Aztecs and Mayas believed the cocoa bean to be a 'gift to man from the gods'. The Mayas also gave the world tomatoes and vanilla.
USA	The Spruce is sacred to the Hopi Indians who also preserve legends of Atlantis.

THE GODS OF AGRICULTURE

The Incas of antiquity were, of course, reputed to be on friendly terms with the gods. Several long-forgotten and even unknown foods to the rest of the world have been found in the Andes and other parts of South America; the tamarillo and potato are examples. Europeans discovered the potato in Peru some 400 years ago, while the tamarillo's origin is also Peru. Recently, even other lost foods such as Quinoa and Kanahua are being 're-discovered' on the Andean Plateau. Were these items 'gifts from the gods'?

Gifts of food from space beings to Earthlings are mentioned in the venerable Stanzas of Dzyan said to be of Atlantean origin. The Chinese book of Y-King records that cosmic visitors taught agriculture to men of Earth. The *Way of Life* by Lao-Tzu speaks of a gentle being who taught agriculture, then flew to the heavens with a team of 'flying dragons'. The Greeks of the fifth century B.C. said the gods had taught agriculture to the people of Attica.

The following also appear to be of mystery origin:

The Mountain Ash or Rowan tree, sacred to the Rosicrucians; the origin of the cotton shrub which once bloomed in divers colours remains undiscovered; the *Nelumbium speciosum* Lotus is sacred while, enigmatically, the olive bough was a peace offering.

A Theosophical belief is that the banana was developed from a melon on the lost continent of Atlantis. The true origin of the banana, plantain, and coconut has not been discovered.

Plato (427–347 B.C.) writing in the *Timaeus*, said that many varieties of flowers and fruit, coconuts, and chestnuts, and several other items, all came from Atlantis.

The ancients believed that the Cedar, Cypress, and Pine were brought from 'paradise' by Seth. Cypresses and Palms were worshipped in Rome, Mexico, and Central America.

WAS THE CABBAGE ARTIFICIALLY DEVELOPED?

An intriguing scientific observation is that certain plants will 'signal' to the Sun for more light which of course is not forthcoming. The cabbage is an example: electronic instruments detect changes in pressure of water vapour built-up on the cabbage leaves which indicates that the cabbage really is signalling for more light. These experiments were first conducted by Soviet phytobiologist Karamanov, and reported in 1960 in the USSR science journal *Selskoye Khozyaistvo.*

A cabbage signalling to the Sun for additional light cannot be considered as a natural characteristic. Nature would not create the cabbage to signal for increased light from the Sun, unless the Sun responds, which it does not. Was the original ancestor of the modern cabbage *artificially developed* under conditions of *artificial light* which could be varied constantly, according to the requirements of the developing vegetable, thus instilling in its cellular memory the automatic stimulus to signal for more light as needed?

Was the first cabbage artificially developed on Lemuria, Mu, Atlantis, etc., or . . . another planet?

Do not underestimate the complexity of a plant cell; it is no less complex than that of an animal or man, and like Man, it has its own DNA molecules which carry construction and memory 'blueprints' that do not usually deviate unless exposed to radiation/LASERS, or another damaging factor.

An item in *Science News*, 22nd July 1967, may support my own hypothesis. Professor B. S. Moshchov, head of the Institute of Light Physiology at Leningrad, reported the development of a new hybrid vegetable – a cross between a cabbage and a radish; it was classified as the world's 'fastest growing foodstuff'. How does this relate with my own theory? The notable point is that this new hybrid was exclusively grown and developed under *artificial light*, proving that the cabbage responds exceedingly well to artificial light conditions!

If twentieth-century scientists can produce a cabbage hybrid under artificial light, who can say that scientists of antiquity were not able to develop the *first* cabbage under artificial light? In 1970 Soviet scientists reported that photosynthesis processes in plants can be speeded-up by growing them in an electrical field. This further supports my own supposition.

WAS MANNA OF EXTRA-TERRESTRIAL ORIGIN?

What is 'Manna' described in the Bible? A tree (*Fraxinus ornus*) growing in Italy, Sicily and Hungary, releases an edible sap called Manna; another Manna is the lichen *Lecanora esculenta* which grows in Africa and Asia, but neither of these substances is the Manna which the Israelites received during their Exodus from Egypt to Palestine!

The 'bread' which the 'Lord' who 'appeared in the cloud' (UFO?) gave to the 'children of Israel', was described as 'like coriander seed, white; and the taste of it was like wafers made with honey'. *Exodus 16:31.*

The House of Israel named the wafers 'Manna'. An

important observation can be noted in *Behind the Flying Saucers* by the late Frank Scully, who referred to the food supply found on a UFO which reputedly crashed in the USA in the late 1940s. Scully said that there were 'wafers' of a very condensed food which swelled considerably when soaked in water, and that guinea pigs 'thrived on it'.

Scully's book has not been disproved. Did UFO pilots give the same type of wafers as food to the Israelites three and one half thousand years ago?

POSSIBLY WHEAT AND CORN WERE BROUGHT FROM CANIS MAJOR

H. P. Blavatsky (1831–91) who founded the Theosophical Society in 1875, and wrote several learned works, mentions in *The Secret Doctrine* (1888) her research into Egyptian history, in which it is recorded that the gifts of wheat and corn were brought to ancient Egypt by the deities Isis and Osiris who descended from the skies.

During the reign of Ptolemy II (Philadelphus, 283–246 B.C., a profound work was written by the erudite Egyptian priest and historian Manetho of Sebennytus. The first book of Manetho states that Isis and Osiris reigned 35 years, and Horus, the 'divine' son of Isis, reigned 23 years.

Also recorded in Egyptian legends is that Isis said she was the *first person* to bring the gifts of wheat and corn, and that she came from the 'Constellation of the Dog', i.e. Canis Major, 8.7 light years away.

Were Isis and Osiris space beings from a planet in Canis Major?

Isis was reputed to have instigated the introduction of beer brewing in Egypt of old. It is interesting to note, in this connection, that Rameses III consecrated 466,303 containers of beer to the gods.

Did Isis bring wheat and corn from the Canis Major constellation? An Earth origin for wheat has not been discovered.

The Rosicrucian texts refer to the 'Lords of Mercury' and 'Lords of Venus' who came to help and teach Earthman. Mercury, 'a great teacher of mankind', was identified with Sirius in Canis Major, as was Isis and the eleventh Theban king.

The Koran records that the archangel Gabriel came from the constellation of the 'Dog-Star'. Mohammed was reputed to have been taken to the 'seventh heaven' (Saturn?) by Gabriel, on a flying white steed (a spaceship?) with 600 wings, about A.D. 597.

Diodorus Siculus or Diodorus of Sicily, the celebrated historian of Egyptian antiquities, said that gods (spacemen?) reigned in Egypt for a period of just under 18,000 years, and that the last god to reign was Horus, the 'divine' son of Isis.

What secrets possibly relating to spacemen overlords still remain under the sands of Egypt? Ancient Egyptian priests catalogued 47 tombs with the mummified bodies of kings but this number have not been located. It is suspected that the Copts of Kubts, Arab-Christian monks, cherish records of the undiscovered tombs, but they remain silent.

The Copts who reside in their holy places on the Libyan Desert border are considered to be the last ethnic remnants of true ancient Egyptians.

Orthodox thought is that modern Man has existed on this planet but a short period, though, curiously, Egyptian astronomical records date back 87,000 years, while the Hindu Zodiac goes back 850,000 years!

WRITTEN PROOF OF UFOs IN OLD EGYPT

The 'winged disc' of ancient Egypt, assumed to be a symbolic representation of the Sun traversing the heavens, could be something *very* different! It seems to bear strong similarity to modern-day UFOs! The Egyptian *Book of the Dead*, written between 1600 and 900 B.C., states that the great 'flying discs' were actually made (or 'forged') in Edfu, a town in Upper Egypt, close to the Nile on the West Bank.

It is recorded that:

> 'from the height of heaven, Heru-Behutet was able to see his father's enemies, and he chased them in the form of the great winged disc.'

The following excerpts are from the *Book of the Dead*; all these papyrus sheets are kept in the British Museum.

The *Papyrus of Nebseni*, No 9900, sheet 6, states:

> 'The white (or shining) eye of Horus cometh. The brilliant eye of Horus cometh. It cometh in peace ...' (a UFO?)

Another interesting sheet is the *Papyrus of Nu*, No 10,477, sheet 8, which says:

> 'There is a serpent on the brow of that mountain, and he measureth thirty cubits in length; the first eight cubits of his length are covered with flints and with shining metal plates.'

According to legend, Osiris knew the name of this mysterious 'serpent'. Was it a UFO? A Memphis cubit was approximately 52 cm long, which would make the 'serpent' about 15 m 70 cm to 16 m long. Any relation between this metallic 'serpent' and the 'serpent' space beings is probably purely coincidental, and of no real bearing on my Chapter 1 hypothesis.

The *Papyrus of Nu*, sheets 9, 17, and 20, in four excerpts, records that:

> 'I have alighted like the hawk by the divine clouds, and by the great dew. I have journeyed from the earth to the heaven.'

Were the 'divine clouds' UFOs?

The second excerpt mentions Osiris in his sacred 'boat':

> '... the boat saileth round about in heaven, rising like the sun in the darkness.'

'Rising like the sun in the darkness' – a glowing UFO?

Again, referring to Osiris in his sacred 'boat', sheets 17 and 20 state:

> '... he cometh upon the flame of thy boat.' 'I have gone down to the earth in the two great boats.'

Were the 'boats' spaceships?

Not only wheat and corn, but also red and white barley were apparently brought to Egypt by Isis from the Canis Major constellation. Referring to these 'gifts' in the *Papyrus of Nu*, No. 10,477, sheet 28, Isis remarked that:

> '... my bread is made of white barley, and my ale is made of red barley; and behold the Sektet boat and the Atet boat have brought these things and have laid the gifts of the lands upon the altar of the souls of Annu. Hymns of praise be to thee, O Ur-arit-s, as thou travellest through heaven ...'

AEROPLANES AT SAQQARA?

An 1898 discovery at Saqqara, Egypt, was a Sycamore wood carving of, according to Dr. Khalil Messiha who studied it at Cairo Museum in 1969, a glider model with a 1 m 13 cm span reverse dihedral wing and a vertical tail fin! Dr. Messiha believes this model from the third–fourth century B.C. may be a replica of a full-size aeroplane, of which remnants may still be found!

THE CHEPHREN PYRAMID EMITS MYSTERY RAYS

Some years ago, the famous American clairvoyant Edgar Cayce said there are secret chambers in the Great Pyramid containing records of Atlantis, placed there by Atlantean survivors after fleeing to Egypt during the final submergence. Cayce said these records would probably be located this century. Air travel, atomic and solar power control, submarines, and electronics, were sciences in which the Atlanteans were proficient according to Cayce. He also remarked that Bimini island – really two islands – off Miami, is an Atlantean mountain peak.

The Chephren Pyramid constructed between 2780 and 2280 B.C. at Gizeh near Cairo has in recent years become the focal point of attention to attempt to locate hidden chambers. The largest pyramid is Cheops, followed by Chephren, then Mycerinus.

Luis Alvarez, 1968 Nobel prize-winner for physics,

and director of the University of California's Lawrence Radiation Laboratory, developed the method used on Chephren. Cosmic rays or 'muons',* i.e. sub-atomic particles, arrive on Earth from space, pass naturally through the pyramid and are then registered in a spark chamber before finally being recorded on magnetic tapes. The IBM-1130 computer at Ein Shams University in Cairo was programmed to analyse millions of sparks, then produce images. If the pyramid contains hollow areas more cosmic rays pass through, and dark areas show in the computer pictures.

Since 1966 when the cosmic ray recorder was installed in the base of the pyramid, thousands of man hours and a cost exceeding $1,000,000 have gone into this project. During one period the recorder was worked continually 24 hours a day for over a year. After all this time, money and work, positive results and discoveries should have been made, but something very, very strange was happening; *an unknown mystery power or force in the pyramid was scrambling the magnetic tapes* in a very unpredictable way!

The cosmic ray recorder functioned to perfection. This was proved, but cosmic ray measurements of the same area over a two-day period revealed entirely different computer pictures when they should of course have been identical!

Dr. Amr Gohed at Ein Shams, in charge of installation of the cosmic ray recorder, said:

> 'It defies all the known laws of science and electronics ... this is scientifically impossible ...
>
> 'Either the geometry of the pyramid is in substantial error which would affect our readings, or there is a mystery beyond explanation – call it what you will, occultism, the curse of the Pharaohs, sorcery or magic, there is some force that defies the laws of science at work in the pyramid.'

Possibly very significant in this connection is that certain ancient Egyptian legends say that Chephren was the 'guardian' of the fourth-dynasty tombs. Has modern

* mu-meson; heavy electron.

science accidentally discovered a protective mystery force in the pyramid, thereby validating these ancient legends?

Did spacemen build the pyramids, or were the builders surviving Atlanteans? The following chapter examines Atlantean 'force fields'.

If Atlanteans were so advanced scientifically, it does of course seem probable that they, themselves, were very advanced space beings who had colonized Atlantis. Enigmatically, a curious name was given to Atlantis by an ancient Greek philosopher; he called it 'the Saturnian Continent'.

3. HAVE ATLANTEAN MAMMOTH AND MASTODON BONES BEEN DISCOVERED?

'They are ill discoverers that think there is no land when they can see nothing but sea.'

(Francis Bacon, 1561–1626.)

Plato remarked that a portion of Atlantis abounded in hot and cold springs, and today, in this very same area, we find that hot sulphur and cold springs do abound in the Azores mountains! Discoveries since 1930 have almost proved the past existence of the Atlantean continent, and have placed its site precisely where it was reputed to have been situated.

The Spanish Conquistadore Bishop Diego de Landa put to the flame in 1562 priceless irreplaceable Mayan manuscripts; what was inscribed in these rare works we may never know. The only extant Mayan ms. fragments are preserved in the museums of Dresden, Paris, and Madrid; but the most significant Mayan work, a copy, is the *Book of Chilam Balam* from Chumayel, which relates to the downfall of Atlantis. This work, partly translated in 1930 by the Brazilian philologist and scholar A. M. Bolio, recounts that the catastrophe which occurred 'during the Eleventh Ahau Catoun' was accompanied by a fiery holocaust and a violent upheaval which ended quickly when the sinking land was inundated by the sea. A strange passage in this manuscript refers to the 'Great Serpent' which was 'ravished from the heavens' during the disaster. One can only relate this to the serpent space beings, probably from Ophiuchus.

Over the years fossils, plants, sand, rocks and lava, which once existed on dry land only a few thousand years ago, have been dredged up from close to Cape Cod, the mid-Atlantic ridge, and north of the Azores; while in

1964 the Woods–Hole Oceanographic Institution, the Navy Hydrographic Office and the Hudson Laboratories of the US reported that a 'chain' of extinct submerged volcanoes was discovered ranging from Bermuda to the New England coast. This discovery supports the 1898 find of atmospherically-solidified lava dredged-up from in this vicinity!

Also in 1964, Professor Georgiy Lindberg, of the Zoological Institute of the Academy of Science of the Soviet Union, stated through *Tass*:

> 'The hypothesis that there is a North Atlantic continent, presently submerged beneath 4,500 to 5,000 metres of water, is confirmed by new findings.'

Science Newsletter of 3rd December 1966, stated that scientists had discovered a large quantity of mammoth and mastodon bones and teeth on the Atlantic Continental Shelf, in some places only 81 metres under the surface of the ocean.

Dr. Frank C. Whitmore, palaeontologist with the US Geological Survey and the Woods–Hole Oceanographic Institution, said:

> 'Records of the rare bones indicate that the shelf was once dry land . . . mastodons and mammoths, forebears of today's elephants, must have roamed these areas some 11,000 to 25,000 years ago.'

This discovery confirms Plato, when he referred to the downfall of Atlantis nearly 11,600 years ago (or the last portion to go under).

THE REMARKABLE EVIDENCE NEAR BIMINI

The most startling evidence of all, apparently confirming the past existence of Atlantis, was only discovered in the late 1960s. Edgar Cayce's statement that Bimini is an Atlantean mountain peak, has, it seems, been confirmed!

The *Miami Herald* of April 1956 reported that massive marble columns were discovered at a 21-m depth off Bimini, but more was to come. Cayce first mentioned Atlantis in 1924, and in 1933 he said that 'a fine example of an Atlantean temple' would be discovered near Bimini in 1968. Again, in 1940, he said: 'the first portions of Atlantis' would 'rise' from the sea in 1968–69. This is precisely what happened! On 16th August 1968, the remains of an ancient submerged building – undoubtedly the temple – was sighted from an aircraft. The temple remains in shallow water, less than 90 m offshore from the north tip of Andros Island, east of the Great Bahama Bank were found to measure 30 m × 18 m, with walls 90 cm thick. The remains had risen 60 cm above the sea bed, but obviously went down much further.

According to Cayce, the 'temple' was originally constructed with a special dome of crystals to capture the Sun's energy, but the dome no longer remains, or perhaps excavation will discover the remnants?

These discoveries are probably the first *real* step to proving the existence of the legendary Atlantis. Illogical archaeological theories have recently been forwarded to suggest that an insignificant land mass in a different location was the real Atlantis, but in view of the Bahama discoveries, it now seems that Atlantis was in the very location as recorded in countless ancient works. Solon, Plato, Donnelly – and others – were not fools.

Edgar Cayce could not have known about the archaeological discoveries near Bimini, as they were many years after his time (1877–1945). One other significant point mentioned by Cayce, strikingly precise in his prognostications, was that the Atlanteans were the cause of their own destruction through nuclear and other forces.

From preliminary archaeological study of the building, or temple, ruins, it was established that the precision limestone-block method of construction is 'Cyclopean', the oldest method of building known to ancient Man, and *very importantly*, no records exist of any advanced civilization in the area! An archaeologist referred to the find as one of the:

> 'most exciting and disturbing discoveries of the century. It is definitely man made.'

The discovery is of such magnitude that the giant

scientific institution, the North American Rockwell Corporation, fought for and won, in 1971, exclusive exploration rights from the Bahamaian government. Other discoveries since – evidently all dated at about 12,000 years in age by the carbon-14 method – include marble and ceramic carvings, the remains of a pyramid, ten or more other buildings, a 542-m-long stone 'causeway', with the stones embedded in cement, 3 fathoms deep off Bimini's north shore, and segments of a tremendous wall which may encircle the whole of the Bahamas. This discovery may validate ancient drawings of Atlantis showing vast areas encircled by several great walls. Two divers on the project, Robert Ferro and Michael Grumley, mentioned in their book that the stone blocks did not belong to the localized strata but were quarried thousands of miles away in the Andes! One little stone with an almost white underside emitted a metallic sound when struck; this cannot be explained. A total of 50 fluted marble columns have now been found in the Caribbean.

GREEN ATLANTEAN RAYS?

A subject of equal fascination to the Bimini–Andros islands discoveries, is the 'Twilight Zone' or 'Triangle of Death', which is a triangular patch of Atlantic ocean between San Juan in Puerto Rico, Miami, and Bermuda. For about one hundred documented years, ships and sailors have vanished without trace in this zone, and in more recent years even planes and crews. Similar patches of ocean are known to exist in other parts of the world which are equally as dangerous, including an area in the Pacific where Lemuria or Mu was once believed to exist!

On 5th December 1945, five small Avenger US Navy torpedo bombers flying from Fort Lauderdale to the Bahamas, vanished in this area after reporting a *green light* under the sea. Approaching landfall in the Bahamas Columbus also noted 'strange lights' and duly reported the phenomena in the ship's log.

A US Air Force bomber pilot fortunate to survive a flight over the Triangle of Death, said:

'We were 300 miles from Bermuda on a beautiful clear night when we were suddenly thrown over on to our backs as the ship was hurled about in a most incredible fashion.

'Then we began to dive, and it took the combined strength of myself and my co-pilot to haul the plane back.

'When we eventually levelled out, we were so close to the water that the white caps caused by our slipstream were visible.'

Major ships that have disappeared in this zone over the years, were:

Cyclopes, the US Naval Supply Vessel. *Sandra*, the US freighter. *Marine Sulphur*, the US tanker. *Freya*, the German timber ship.

Ships with strangely vanished crews in this zone, were:

Rubicon (1948), the Cuban freighter. A cabin cruiser owned by US jockey Al Snyder (1948); and others well known.

In recent years at least twelve planes – excluding those already mentioned – have vanished in this area. The zone is considered dangerous, and the triangle is watched by the US Coast Guard, Navy, and Air Force.

What really happens in the Twilight Zone or Triangle of Death? There seems no doubt that an inexplicable power or force in this area can draw planes into the sea and sink ships. There are two significant facets to this enigma. The first is the geometric triangular shape of the zone which can hardly be considered natural, and the second facet are discoveries in these areas evidently proving the past existence of a highly advanced section of Atlantis!

Are ancient Atlantean electronic or magnetic devices, once used to repel invaders, i.e., 'air-boats', UFOs, sea battleships, or submarines, *still operating automatically in this area*? Perhaps drawing power from Earth while deep beneath the sea or sea bed, and emitting lethal disintegrating rays which destroy planes, ships, etc.?

Is the triangular shape of the zone an electronic or

magnetic screen or curtain, once used to protect an important area of Atlantis from hostile invaders? Are three ancient electronic or magnetic devices at or near the triangle points, deep under the sea or sea bed, still forming a protective curtain no longer needed?

Did such devices survive the downfall of Atlantis? Scientists cannot offer a convincing solution about the mystery power!

IS AN ATLANTEAN ISLAND SHOWN ON THE PIRÎ REIS MAP?

Much discussed in UFO publications since the 1950s, the controversial Pirî Reis map, copied by the esteemed Turkish Admiral Pirî Reis in 1513 from ancient Greek charts of an even earlier origin, shows the surface of the world from a great height; seemingly originally photographed by spacemen as it was then, several thousand years ago.

On studying my own Library of Congress photo-copy of the Pirî Reis map, I noticed an island – not mentioned or noticed by other UFO authors in this context – portrayed between Brazil and the West Coast of Africa. Needless to say, this particular island does not exist today.

Comparing the Pirî Reis map with a Theosophical map of the ancient world, I was intrigued to see this same island, known in Theosophy as 'Daitya', a large Atlantean island, existing at precisely the same geographic location; but on the Pirî Reis map, Atlantis itself seemed to have already sunk.

If in fact the Pirî Reis map does portray an Atlantean island then the original of this map may be dated somewhere in the vicinity of 11,500 to 12,000 years ago, but not likely to be very much more recent than 11,500 to 12,000 years ago, as Atlantis reputedly sank about this time.

INDIAN VIMANAS MAY HAVE BEEN CONSTRUCTED FROM ATLANTEAN BLUEPRINTS

'Aryavarta' from Sanskrit – 'the language of the gods' – in its literal translation, means: Abode, or land of the Aryans; classified by scholars as India of old, but in reality Aryavarta was a tract of land between the Himālaya and Vindhya ranges.

In *The Secret Doctrine* Madame Blavatsky related that the early Aryans (Atlanteans?) learned astronomy, and many other skilled sciences, from an even earlier race – apparently an Earth race – who also taught them 'Viwan-Vidya' or aeronautics. Was this earlier race Atlanteans – once space beings themselves?

Two fascinating hoary astronomical works, briefly described here, have survived from archaic India, but what is even more intellectually stimulating is the precise accuracy of these codices, recorded by a people who were not supposed to possess advanced astronomical instruments!

Kaushitaki Brahmana, a Hindu work dated approximately 3100 B.C., contains unusually precise and advanced astronomical knowledge; even older is the Vedic Taittiríya-Aranyaka, which also preserves records of interesting astronomical observations. Vedic (from 'Vid' – 'to know') was an even more ancient language than Sanskrit, and dated back to about 3000 B.C.

Both these very old codices validate wisdom recorded by Madame Blavatsky. It seems that early Aryans *did* learn astronomy from an earlier race, or races.

The knowledge for constructing and flying aerial vehicles described by Madame Blavatsky really refers to the ancient Indian Vimanas, usually represented as boat-shaped, jet-powered vertical-take-off craft of limited flying capabilities. These were reputed to have been identical to the earlier VTOL Atlantean air-boats.

Vimanas, and the narrative from the Hindu epic *Rāmāyana* of Rāma's wife Sītā being abducted from India by the evil king Rāvana and flown back to

Sri Lanka in a Vimana, have been thoroughly analysed by many UFO authors. However, in this connection I can offer two original gleanings.

Some proof attesting to the authenticity of these tales are extremely old rock frescoes at Sigiriya in Sri Lanka, thought by a noted Ceylonese archaeologist to have been painted by an artist of the Court of Rāvana, thus indicating the existence of Rāvana as an historical fact. Another find was made by the Soviet researcher Aleksandr Kazantsev in the Leningrad Hermitage Museum. Kazantsev discovered two enchanting little Etruscan cameos some 3,000 years old, portraying most singular and rare scenes. One priceless cameo showed a boat-shaped craft emitting *rays* or *jets* from the aft section, but *without* oars! Circular appendages along the side of the vessel correspond with the description of jets on Atlantean air-boats in *The Story of Atlantis* by W. Scott-Elliot (1896). Does this cameo depict an Atlantean air-boat or an Indian Vimana? The other cameo was evidently presenting in visual image, cut in ornamental relief, a spaceman wearing a bulky neck-to-foot outfit, while over his head was a large spherical helmet!

4. THE FIRST SPACE BEING SETTLERS IN INDIA LANDED ON THE LONGKAPUR HILL IN THE LOHIT VALLEY

'All this visible universe is not unique in nature, and we must believe that there are, in other regions of space, other worlds, other beings and other men.'

(Lucretius, 99–55 B.C.)

People of India's North-East Frontier preserve many legends relating to space beings arriving on Earth in times long past. Some legends speak of times when spacemen and Earthmen lived together in coexistence; a few spacemen harboured hostile tendencies towards Earthmen, while others were concerned with helping and teaching new sciences to Earthlings. The 'heavenly beings' known to the Hindus as 'Siddhas', were called 'possessors of knowledge' and 'masters of high science'.

A Hindu painting by an unknown eighteenth-century artist, reproduced in the *Sydney Morning Herald*, portrays the sky vehicles of the gods; at least two sky vehicles in the picture resembled modern UFOs.

The Hrussos, or Akas, of the North-East Frontier, say that at one period of pre-history there were no men on Earth, and *we are all descended from space beings who migrated here in antiquity!* They also relate that the first space beings to arrive in India landed on the Longkapur Hill in the Lohit Valley; from there they spread out through India and Tibet.

According to one legend, a certain group of spacemen were lazy and passed their time beer drinking, so missing out on the best land.

The women of Sangtam in the North-East Frontier wear symbolic discs of white shell around their necks.

Are these UFO representations? Not only Egypt, Assyria, Guatemala, Persia and other lands, but also India of 6,000 years ago, possessed symbolic representations of the 'winged disc'!

The Hindu pantheon contains the incredible number of 33,000,000 deities. There seems no doubt that some, at least, were space beings!

The Purānas speak of the Sanakadikas or 'Ancients of the Space Dimensions' while other codex texts refer to the Gandharvas or 'heavenly musicians', 6,333 in number; the 'Solar Race' or Sūrya-Vansa, and the Siddhas already mentioned, who came from the sky between Earth and the Sun, which seems to indicate Venus or Mercury. Did the Siddhas really come from Mercury or Venus?

The Arab astronomers of old named Venus Al-Azza and in 12th-century Europe Venus was known as Lucifer. In July 1964 Soviet scientists announced that a radar probe proved that Venus has 'a hard rocky surface similar to that of Earth . . .'

The Soviet Venus probes: *Venera-5* which descended through the Venusian cloudy layer at 0601 GMT, 16th May, 1969, *Venera-6* which followed one day later at 0605 GMT, 17th May, and *Venera-7* which descended 18.1 km to land on the Venusian surface on 15th December, 1970, all indicated a Venusian environment inhospitable to life as we know it on Earth. But these probe broadcasts were in diametric opposition to innumerable references in historical codices, etc., which *positively state* that Venusians have been visiting Earth in spaceships since great antiquity – at least 16.5 to about 19, or even as far back as 78,000,000 years! The Venus probes – including *Venera-8* – are *still* at an early stage. Perhaps time will tell if there is an unknown factor as yet undiscovered? Also, many Venusian mysteries still remain – the anomalous axis rotation of Venus, or why the upper clouds revolve in opposite direction from the planet in a 96-terrestrial-hour period, the structure and composition of Venusian clouds, and upper atmosphere processes, and why the 'solar wind' plasma 'goes round' the upper atmosphere. Are the clouds artificial*?

Scientists are expeditious in arriving at conclusions for the non-existence of life or life-forms on Venus, but back in 1963 when Vladimir Prokofyev detected oxygen in the Venusian atmosphere by operating a unique type of spectrograph at the Crimean Astro-Physical Observatory, the French newspaper *Le Soir*, 31st August, 1963, said:

> 'Radio Moscow has announced that the planet Venus has an atmosphere containing oxygen and that it will, therefore, be susceptible to being peopled by beings similar to Earthlings. . . .'

This report appeared nearly nine months *after* the US *Mariner 2*'s 'fly-by' on 14th December, 1962, which indicated an inhospitable Venusian environment!

Venus-5 reported a Venusian surface temperature of 530° Celsius at 140 atmospheres, *Venus-6* 400° C at 60 atmospheres, and *Venus-7*, from the surface, a temperature ranging though ±20° C on either side of 474° C.

In 1965 scientists at Baltimore's Johns Hopkins University discovered evidence of water ice in the Venusian clouds. This indicated that Venus' surface temperature could, in actuality, be in the vicinity of 4½° C – a temperature Earthman would be able to tolerate, but cold!

Why did the USSR Venus probe give such high temperature readings? Hopkins scientists believe that earlier (and now later?) 'hot temperature' readings from the Venus probes may be due to a large degree of lightning-like electrical activity taking place in the thick Venusian cloud cover which gave false readings. The Soviet probes gave very high temperature readings for the Venusian surface. If, as thought by Hopkins scientists, a large amount of lightning-like electrical activity may be taking place in the Venusian clouds, could this also affect radar, and radio signals sent through the clouds from the *Venus-5*, *-6*, and *-7* probes in the atmosphere and on the planet? Possibly sending back to Earth *false figures* –

*Inexplicable, is that over a 96-hr period a 20% shift in CO_2 absorption intensity occurs with a 1-km cloud-level change. The phenomenon requires vast planet energy release, but cannot be related with known Venus characteristics.

before, perhaps, destroying the probes' electronic systems – of not only temperature, but also the percentages of atmospheric gases?

Upper ionosphere Earth probes reported that life as we know it cannot exist on our planet, for the probes recorded an Earth temperature of 982° C with *no* oxygen or water vapour! So life on Venus, or other planets, cannot be ruled-out after these experiments. Perhaps, too, the radio data from the Venusian probes, possibly affected by electrical activity in the Venusian cloudy layer, may be still further affected passing through space and Earth's ionosphere? (Or do Venusians – and others – scramble out signals?) Dr. William E. Plummer, a scientist at Johns Hopkins, said:

> 'Like the roof of a greenhouse, the clouds would also reflect a great deal of the sunlight back into space.'

The Hopkins scientists believe that there may be 'large intermediate areas' on Venus, where the temperature is not too extreme, and may 'readily support life'. A summary of findings by Hopkins scientists reported that their discoveries:

> '. . . re-open the question of life on Venus, that many scientists had ruled-out too casually, after previous announcements of very high temperature readings on the planet's surface.'

Both Soviet and US Venus probes have given conflicting readings. The Russian *Venus-4* probes detected no magnetic field around Venus on 18th October, 1967, but *Mariner-5* passing one day later, detected 'definite magnetic activity'.

Although *Venus-4* functioned for 93 minutes in the Venusian atmosphere, its readings of carbon dioxide, nitrogen, oxygen, water vapour, temperature and atmospheric pressure, *all* varied from the 1969–70 readings of the Venusian atmosphere, etc., by *Venus-5*, *-6*, and *-7* – which, again, varied among themselves.

Tass reported in September 1964 that astronomers at Kharkov University Observatory in the Ukraine photographed a 'large dark spot' on Venus. Academician Nikolai Barabashev believes the 'dark spot' was 'a great clearing in the upper layers of the clouds' which enabled astronomers to see, and photograph the surface; but what did they see, and photograph?

While not definitely advocating Mercury as an abode for intelligent beings, on the basis of very slender evidence available, let us now look at Mercury as a (possible) association with the Hindu deities Siddhas – 'possessors of knowledge' – 'masters of high science'.

A 1964 radio-astronomy probe from Parkes, north-west of Sydney, proved that the dark side of Mercury has a comfortable temperature of 16° C. At Pic du Midi, the French Observatory in the Pyrenees, Dr. Dollfus discovered a slight atmosphere around Mercury while he was conducting polarization studies. This was later confirmed by the Soviet astronomer N. A. Kozyrev, who estimated the atmosphere to be 1/10,000th of Earth's atmosphere, and probably composed of atomic hydrogen, which is also a component of Earth's ionosphere.

In 1965, US radio-astronomers operating the 301-m 'dish' at Puerto Rico, discovered that Mercury goes through day–night cycles as do other planets in our solar system. This very same radio-telescope, owned and operated by Cornell University, indicated that Mercury has a denser composition than the Moon, but a similar landscape.

DID MAHENDRA POSSESS A UFO IN THE THIRD CENTURY B.C.?

At Sanchi (population 1,000) between Bhopal and Jhansi in Madhya Pradesh State, Central India, is 'The Great Stupa'; a Buddhist monument begun by Emperor 'Asoka The Great' in the third century B.C., but completed at a later date. To a researcher of extra-terrestrial phenomena, The Great Stupa strongly resembles a UFO. With power coils?

There is an unmistakable and striking resemblance between a photograph of The Great Stupa, and a photograph of a reputed landed UFO at Lossiemouth,

Scotland, taken by Cedric Allingham on 18th February, 1954. The Great Stupa is approximately 2.4 times larger in diameter, and 2.7 times higher, in dimensions, than the Lossiemouth UFO.

The following curious legend relates to Asoka's son Mahendra:

> 'On a full-moon day, with his companions, Mahendra rose into the air from Sanchi, and flying aloft as fly the golden geese, alighted on the peak of the Mihintale mountain in Ceylon.'

On arrival in Sri Lanka (near Anuradhapura) Mahendra converted the king to Buddhism. Was The Great Stupa constructed in the shape of a UFO piloted by Mahendra to Sri Lanka? Was Mahendra's UFO (?) identical, or similar, to the Lossiemouth UFO? Very striking is an antenna-like appendage on top of The Great Stupa, very similar to the cabin antenna on the Lossiemouth UFO!

The Mosque of Omar constructed on the site where Mohammed 'ascended to heaven', also resembles a UFO!

ENIGMAS FROM THE INDUS VALLEY, HIMĀLAYAS, AND DELHI

Five thousand years ago a largely unknown and unidentified people lived in their city Mohenjo-Daro in West Pakistan's Sind Desert. One day the entire population suddenly vanished leaving the city in a perfect state of preservation which can be seen today.

Where did all those people go? An undeciphered 5,000-year-old seal from the Indus Valley shows an embossed figure wearing what may be a bulky space suit and helmet, quite different from other figures with facial features on this seal!

The *Karachi Evening Star* (Pakistan) for 10th October, 1967, reported that one Ruth Reyna, a physicist and physics teacher at the East Punjab University of Chandigarh, submitted a report to NASA in which she claimed that in 3000 B.C., 1,000–1,200 people left Earth from the

Indus Valley by spaceship and travelled to Venus where they formed an 'Indian Colony'. According to Dr. Reyna the reason for leaving Earth was due to imminent catastrophe forecast by the astrologers of the day.

WHO WAS AGASTYA?

Hindus of antiquity worshipped the star Canopus in Argo which they named Agastya-Muni. A very old village in a Himālayan valley is also called Agastya-Muni. Curiously, a Brahmin saint of old, credited with founding Tamil literature, was named Agastya.

Was Agastya a spaceman from the vicinity of Canopus? The ancient Egyptians also venerated Canopus and they too named a place after this star: the city of Canopus or Canobus east of Alexandria. (Any relation to Serapis the Egyptian god of healing?)

DELHI

Delhi is thought by some to be the ancient city Indraprastha, capital of the Pandavas, which is mentioned in the epic *Mahābhārata*. It was reputed to be the city where gods and mortals once walked.

Along the riverside are ancient mounds believed by some archaeologists to contain archaeological secrets of antiquity, previously guarded for hundreds or thousands of years.

What will the mounds disclose when archaeologists begin excavation? Archaeological secrets relating to spacemen?

DO SPACEMEN LIVE FOR HUNDREDS OF YEARS?

Ancient Hindu records state that one year of the gods equals 360 years of Earthman. Does this mean that the gods lived on a planet taking 360 years to complete one orbit of a sun?

Earth takes one terrestrial year or 365 days, 5 hrs. 48

min. 46 sec. to complete one orbit of our sun, and this is known as the 'solar year'. Pluto, our previously considered outer planet takes 248.43 years to complete one orbit of our sun. Another planet beyond Pluto might have an orbital period around our sun of 360 years although some theories suggest that a tenth planet would have an orbital period of 675–6* years and a distance double that of Pluto from our sun.

Did the Hindus of antiquity mean that the gods came from a solar system with a planet taking 360 years to complete one orbit of a sun? What is the real meaning?

Of possible interest in this connection is that the Hindu deity Indra was believed to have come from Swar-Loka, a planet between our sun and the Pole Star.

An alternative possibility to the above planetary orbital theory could be found in Einstein's Special Theory of Relativity, apparently known to certain Oriental, Slavic, and Near East peoples thousands of years ago. Example: if an Earthman undertook a lengthy spacecraft journey into deep space he might return to Earth still young while his descendants had expired through old age. So, Earthlings might age at a metabolic rate more quickly than in space. Or, by terrestrial time measurement perhaps 360 Earth years equals but one (terrestrial) year in deep space?

The *Mahābhārata* recounts that the Indian gods of antiquity lived many hundreds of years; this same characteristic evidently also applied to Earthman.

George Adamski, the late American UFO author, mentioned what may be a possible reason for a reduced life-span of present-day Earthman – i.e. the dispersal of a watery cloudy layer above Earth, and a corresponding increase in the bombardment of primary cosmic rays. The dispersal of the cloudy layer ostensibly caused the Flood described in Judaic records.

A dramatically weakened terrestrial magnetic field would effect an increase in the impingement of primary cosmic rays, but in this context it seems unlikely.

*A new theory suggests 464 years.

Adamski, among other UFO writers, said that Venusians live for hundreds of years, and the reason could partly be the water content of the Venusian cloudy layer which may absorb a percentage of primary cosmic rays. Russian Venus probes have hitherto indicated a cloud layer water content of 4-11 milligrams per litre at 0.6 atmospheres.

What caused the Flood? Why did the cloudy layer disperse? Aborigine legends indicate that the serpent beings may have had something to do with causing the Flood (?). (Refer also to first page of Chapter 3.)

Perhaps Earthman will once again live for hundreds of years? It may be feasible to increase the lifespan up to a point, or a previous point, but we still have to contend with the seeming eventual (ironic) ageing factor of the chemical ingestion of iron making the DNA molecules more brittle and less durable. But perhaps this is not a bodily weakness? A metabolic change may have taken place in the human body due to increased cosmic radiation affecting the hereditary DNA and muscle protein molecules, thereby causing them to absorb excessive iron? But even if this was so, which is unproven, for all the marvellous precision of the human body it is still imperfect, for the article: 'Medicine: The Mystery Why We Grow Old' (*New York Times*, 30th October, 1966) states:

> '. . . the conquest of cancer, heart disease and the like will not lead to a dramatic increase in the life span. Too many weaknesses are built into the human frame to be overcome.'

At present Soviet scientists are engaged in intensive research to attempt to discover how to prolong the life of human beings for 600–700 years. The chief purpose is for future space flights which may take as long as 100 years. It seems if Soviet scientists anticipate 100-year space flights this would mean journeying to alien solar systems. If such intensive research is going into this project indications may be that USSR scientists know more than they release about life in other solar systems. Do they anticipate meeting with other beings?

On 10th November, 1967, Patrick Moore asked Russian Cosmonaut Col. Bykovskiy on *The Sky at Night* (BBC-1 TV) why he had such an intense interest in space. Col. Bykovskiy replied:

> 'Who knows, one might meet out there one of the people from other planets.'

Over 100 scientific centres in the Soviet Union are at present engaged in life-prolongation research for the purpose of long space flights. *The Youth Doctors*, written by former *Newsweek* correspondent Patrick McGrady, who as a correspondent in Moscow was able to interview several scientists, quotes an interview with Lev Vladimirovich Komarov, head of the life-prolongation research in the USSR. Referring to the ageing question, Komarov said:

> 'This problem should be resolved with even more resolution than that devoted to the solving of the A-bomb problem a while back, or the current problem of conquering the cosmos.
>
> 'Every year's delay means that some of us will lose decades ... others perhaps even hundreds of years.'

If space beings live hundreds of years it would be a definite advantage if they travel from one solar system to another!

US scientists have experimented with Yoga for its possible use by astronauts on long space flights. Did the Indians of old learn Yoga from spaceship pilots? Perhaps there is much truth in the old adage: 'there is nothing new under the sun'?

A popular science-fiction theme for many years has been 'teleportation' where the entire molecular structure of spaceships and crews is broken down electronically, instantaneously transmitted over vast distances and reassembled at the receiving end. It is not possible to say at this present time whether such a method of space travel would be practicable for Earthlings, but the possibility should not be dismissed that this method of space travel may be in use by extra-terrestrials when travelling from one solar system to another.

This means of space travel may sound fantastic but it is not considered so by many scientists on this planet, for secret research is being undertaken by US and Soviet scientists into a system of space travel called: 'time-stopped molecular decomposition-recomposition'. This means that a spaceship and crew would be broken down without harm in 'pure energy components' into an 'electrodynamic bio-pack'. The bio-pack would be launched into space on a radio beam which would then be directed towards a star with a solar system considered able to support life. Hundreds or thousands of years later when the bio-pack arrived in the gravitational field of the selected star, the spaceship and crew would hopefully be 'reconstituted'.

Due to extensive analysis by other UFO authors I have not covered in this largely Indian chapter the pristine Hindu nuclear war legends. Although in the following chapter these legends are very briefly mentioned, the reader will find that much more than legends only a few thousand years old survive to indicate ancient atomic warfare. Outlined in the next chapter is actual physical evidence still in existence to indicate or prove that raging nuclear holocausts took place on Earth many, many millions of years ago.

5. THE DINOSAURS AND OTHER REPTILES DIED IN A NUCLEAR HOLOCAUST 60–70,000,000 YEARS AGO

'Why is Agni burning all these creatures? Has the destruction of the world begun?'

(From the *Khandava-Daha-Parva* of the *Mahābhārata*.)

While not the most ancient section of this chapter, I am featuring the dinosaur mystery first, due to its very extraordinary nature. The 'great dying' – one of the world's greatest palaeontological engimas – *how*, and *why* did the dinsoaurs 'suddenly' become extinct?

Dinosaurs reigned supreme for 140,000,000 years, then, 'suddenly' disappeared altogether, in the borderline between the Cretaceous and Tertiary periods, 60–70,000,000 years ago!

An 1866 discovery, apparently almost conclusively proving that spacemen were either visiting, or dwelling on Earth during this approximate period, was of a tiny block of lathe-machined meteoritic polished steel, found by the naturalist Gurlt, in a Tertiary – Miocene, Middle Tertiary – coal stratum in the Austrian alps. The 'man-made' steel cube, 12–26,000,000 years old, with a precise groove cut around its middle, measures 70 × 70 × 50 mm and weighs 785 grams. The full report appeared in *Nature*, 11th November, 1886, p. 36.

One hundred years of intensive geological research *have not* disclosed why most of the dinosaurs 'suddenly' perished, strangely, not just as a particular species, but at the same time, and along with, pterosaurs – 'winged dragons' – flying reptiles, and the sea reptiles – ichthyosaurs, mosasaurs, and plesiosaurs.

The one point established with certainty, is that the

'great dying' was largely a 'sudden' violent event, which took place all over the world at the same time! *None* of these creatures died naturally!

The dinosaurs lived in what are now Argentina, Australia, Belgium, Brazil, Canada, England, France, Mongolia, Morocco, Portugal, and the USA, and most, at the same time, came to the same violent end! That is, they were heavily irradiated with intense radiation, scorched with intense heat and immediately buried in astronomical numbers under extremely violent catastrophic conditions! As already mentioned in the 'Introduction', remaining dinosaurs gradually died out, after the catastrophe, seemingly due to genetic mutation caused by intense radiation, which is, of course, the major cause of mutation.

The Russian writer Vladimir Komarov said: 'Who destroyed the dinosaurs?' Who, indeed?

In the early 1960s Soviet scientists developed a new theory as to why the dinosaurs 'suddenly' perished. This new theory is the *only one* that has, with subsequent research, been supported by actual discoveries!

The Russian scientists believe that the dinosaurs disappeared during a period of intense radioactivity. Now the *only* natural process creating such high-energy-level radiation comparable with the dinosaur extinction theory, would have to have come from a close Supernova outburst. A Supernova is an unstable star which suddenly enlarges, casts off its gas shell, and releases enormous amounts of lethal primary cosmic rays, i.e. super-fast-moving electrons and other charged particles, mostly hydrogen protons stripped from atoms. If the Supernova is no more than 25 light years away in radius from our sun, the cosmic ray intensity may substantially increase while spreading towards our solar system; but if too far away, Supernova cosmic rays would scatter in space. However, the possibility of Supernova outbursts within a 25-light-year radius of our sun, is considered to be exceedingly rare. Soviet computer calculations indicate that only two (could) have occurred during the entire period from when life first appeared on Earth to the present day!

The only other consideration for a natural rise in radioactivity, aside from cosmic causes, would be the possibility of radioactive rock emitted from the depths of the earth during mountain-building, and volcanic processes. We will shortly see that not one, but three factors *together*, were associated with the dinosaur extinction. At first thought, the terrestrial possibility just mentioned may supply the answers, but terrestrial rock *could not* possibly supply such intensive high-energy-level radiation *comparable* with a Supernova outburst, to cause mass extinction of giant creatures world-wide at the same time! It is also *extremely* unlikely that *sudden, intense, world-wide* mountain-building and volcanic activity took place at the same time, for *most* of the dinosaurs did die *suddenly, world-wide,* and *obviously* at the same time; this is proved by the fossil record! The radioactive rock theory is unlikely and improbable!

Tests for residual radioactivity in dinosaur and other reptilian bones and fossil remains in the Palaeontological Museum of the Academy of Sciences in the USSR, undertaken by engineer V. Bogoslovsky in collaboration with scientists from Moscow University's Institute of Nuclear Physics, disclosed *extraordinary facts*! Referring to the bones from the mass extinction period 60–70,000,000 years ago, Bogoslovsky said they possessed 'an *exceptionally high* radioactivity'! (my italics).

It has now been established that intense radioactivity was a *major factor* in the dinosaur extinction enigma. Another discovery seemingly in this respect, was dinosaur eggs found in Outer Mongolia in 1922, which in the past had been developing normally, but had 'suddenly' been halted in their development!

Referring to these fossilized eggs, Edwin Colbert remarked in his book *Dinosaurs: Their Discovery and Their World* that:

> 'Some of these dinosaur eggs never hatched. What prevented their development as they lay buried in their sandy crypts is a puzzle, all we know is that no little dinosaurs came out of eggs . . . in a few of the

Mongolian eggs . . . are traces of fossilized embryonic bone, an indication that development had at least gone on for some time before hatching of the eggs was interrupted.'

This discovery could, of course, indicate later production of dinosaur eggs, after the catastrophe – a later period where the earlier radiation had already established genetic mutations – but something *violent* happened in Outer Mongolia. Were the eggs preserved by sudden radiation, or is this an indication of degenerative extinction caused by genetic mutation?

Completely-preserved dinosaur's eggs have been found the world over, along with baby to adult dinosaur skeletons. This of course indicates sudden radiation!

Although the sea reptiles already mentioned perished along with the dinosaurs, some *very deep* sea creatures (and creatures in caves) escaped sudden death and have survived to this day. This indicates that the extreme depth of water, and density of rock, protected them from radiation!

The major role of intense radioactivity in exterminating the dinosaurs has now been established, but what deals a blow to the Supernova theory are two other factors also world-wide. The first is evidence of *intense heat* from an (overhead?) source! Some dinosaur footprints were made in mudflats covered by shallow water, which would have obliterated the tracks, but these tracks were baked into stone and cracked into polygonal shapes, which indicates *rapid vaporization* of the water and dinosaurs, thereby preserving the tracks. Significantly, rarely have dinosaur skeletal bony remains been found with the tracks!

The second factor mentioned, is that buried deposits of dinosaur bones found the *world over* show unmistakable evidence of violent physical catastrophe and *instant* burial; many of these bone deposits are a complete tangle of fragments. Pointedly, skeletal remains of other reptilian life-forms were *also* found with the dinosaur bones!

So . . . there are now *three* proven factors in the extermination of the dinosaurs – *intense radioactivity, intense heat, and massive physical catastrophe!* The implications seem obvious; *only* nuclear explosions can explain these factors together!

Referring to a 1947 dinosaur fossil find in north-west New Mexico, Edwin Colbert said in his book *Men and Dinosaurs:*

> 'They were found in the greatest profusion, piled on top of one another, with heads and tails and feet and legs often inextricably mixed in a jackstraw of bones.'

He further remarked:

> 'It would appear that some local catastrophe had overtaken these dinosaurs, so that they all died together and were buried together.'

But the dinosaur catastrophe was *not* local! Enormous deposits of fragmented dinosaur bones were found world-wide – in Canada, Tanzania-Africa, Bernissart-Belgium – while it was quite obvious that dinosaur fossil deposits in Outer Mongolia showed *unmistakable evidence of instant burial*; and *also where radiation was evident!*

Did spacemen battle in a world-wide atomic war 60–70,000,000 years ago, which destroyed the dinosaurs and other creatures? If so, why? *Was the battle for the possession of Earth?* Or did spacemen *deliberately* exterminate the dinosaurs, diplodocuses, pterosaurs, mosasaurs, ichthyosaurs, and plesiosaurs with atomic weapons to pave the way for the eventual habitation of Earth by Man?

A fascinating Zulu legend handed down through the centuries, refers to 'metal monsters' (spaceships?) which once destroyed the world; remember that fragmented dinosaur fossil remains were found in Africa!

Does the Bible describe an atomic war over and on Earth – possibly in the dinosaur extinction period?

Rev. 12:3, 7, 8, and *9* mentions a 'dragon' (spaceship?) in 'heaven', and a 'war' between Michael the archangel 'and his angels' 'against the dragon . . . and his angels' (or spacemen?):

> 'And there appeared another wonder in heaven;

and behold a great red dragon . . . And there was war in heaven: Michael and his angels fought against the dragon; and the dragon fought and his angels, And prevailed not; neither was their place found any more in heaven. And the great dragon was cast out . . .'

The commander of the 'dragon' – or spaceship? – was 'called the Devil and Satan, which deceiveth the whole world: he was cast out into the earth, and his angels were cast out with him'.

Was the 'Devil' or 'Satan' (a 'serpent' being?) in reality an 'evil angel' or evil spaceman, who attempted to gain control of Earth, only to be defeated?

Flying red dragons are depicted in Chinese mythology as: 'a thousand feet long' which flashed 'fire' and 'lightning' from the eyes, tongue, and mane. There may be more than just a superficial similarity between the *Revelation* account and Chinese mythology!

The *Revelation* description seems to indicate what could be a take-over attempt to gain control of Earth! If this was so, when did it happen – 60–70,000,000 years ago? As already proved, Biblical chronology for certain events of great antiquity is extremely inaccurate.

If deposits of uranium-rich pitchblende ore are natural (fuel or waste products associated with prehistoric spaceships or nuclear devastation?) a possible way for spacemen of antiquity to annihilate extensive areas of land without utilizing atomic bombs, etc., or nuclear missiles, might be to locate uranium deposits then direct *intense narrow-band neutron ray* beams at these deposits, so releasing *enormous quantities* of nuclear energy!

Why were dinosaur tracks, but no bones, found in uranium-rich Arizona desert country? Were the dinosaurs vaporized? And why is this area a desert?

Early theories of climatic change, epidemics, food shortages, and reptilian struggles for existence, are now thought to be highly improbable as answers for the dinosaur extinction.

Some insignificant mammals mysteriously appeared before the dinosaurs and other creatures were destroyed, and the theory that these mammals contributed to the dinosaur extinction by eating their eggs is also now thought of as most improbable. The three factors outlined in this chapter – intense radiation, intense heat, and massive world-wide physical catastrophe, all in the same time period, and apparently at the same time – interrelated – must disturb orthodox palaeontologists and geologists, who while believing in a natural explanation for the dinosaur extinction, have not been able to find one!

One of the strangest enigmas is that *after* the dinosaur extinction, a *diversity* of placental mammals 'suddenly' appeared on Earth. All these new arrivals dealt a blow to the Theory of Evolution, as the fossil records show a 'sudden' appearance of the new mammals in a very short period of time, while the Theory of Evolution requires *long periods* of *extremely gradual* change to support this hypothesis!

Also, after the dinosaur extinction, angiospermous plants appeared, not in the familiar small variety, but in enormous and diverse numbers and types: grasses, cereals, fruits, vegetables, nuts, and flowers. And very shortly Earth was covered with deciduous forests!

If spacemen battled for the possession of Earth 60–70,000,000 years ago, or destroyed the dinosaurs for eventual habitation of Earth by human beings, did the winning team – or colonizing spacemen – repopulate and replenish Earth with mammals and flora? Even radiation-induced genetic mutation would not change reptiles or simple mammals into mammals of wide variation, nor diversify angiospermous plants into such a bewildering array!

Let us now look at the next stage of ancient atomic devastation; the mysterious glassy stones found the world over – tektites – believed by orthodox science to have a natural origin, but certain characteristics associated with tektites cannot be associated with natural origins, for the prominent Soviet physicist Matest Agrest remarked on the orthodox line of thought for the origin of tektites when he said:

> 'These and other hypotheses, however, do not explain all the features of tektites, and their origin con-

tinues to be a puzzle ... in short, tektites, particularly the so-called Libyan glass, may well prove to be marks left by reconnaissance missiles or by the braking and acceleration of a rocket-load of visitors from some remote part of the universe. The prowess and intelligence of the astronauts could well merit them the status of superior beings, "sons of God".'

(Excerpt from thesis: *Astronauts of Yore* by Matest Agrest; translated from the Russian by D. Skvirsky and V. Talmi.)

Certain other factors associated with tektites indicate that photon rockets, as theorized by Agrest, may not have been responsible for forming tektites. Although Agrest forwarded a bold hypothesis for an orthodox physicist, the following factors suggest that tektites may, in reality, have been formed by atomic explosions:

(A) An unusual crater in the Libyan Desert, apparently linked with the Libyan Desert tektites.

(B) Radioactive isotopes contained in some tektites.

(C) A reversal of Earth's magnetic field almost three-quarters of a million years ago evidently associated with tektites.

(D) Tektites approximating in age the time period when the dinosaurs died.

POSSIBLE TEKTITE/NUCLEAR RELATIONSHIP

Tektites, from Gk. 'Tektos' or 'melted rock'. The first tektites discovered were found in Czechoslovakia about 1864 and are dated at 13,000,000 years.

Tektites are in many different shapes and sizes, ranging from tiny glassy stones up to the size of a grapefruit.

Many have been found in Australia where their hitherto dated ages range from 5,000,* 500,000 to 1,000,000 years. One area is the Kimberley Ranges where the inscrutable mouthless beings once visited! In south-east Asia they are dated the same age as in Australia. The world's oldest tektites are found in Georgia and Texas where the dinosaurs certainly roamed. Some of these tektites may be more than 45,000,000 years old.

The following are major tektite sites: Bohemia, Borneo, the Dead Sea area, Indo-China, Iraq, the Ivory Coast in West Africa, Java, Lebanon, Libya, Mexico, Peru, Philippines.

Tektites are known under some of the following names: Australites (Australia); black-green diamonds (USA); Indochinites (Indo-China); Philippinites (Philippines); Libyan Desert Glass. American Red Indians once made them into arrow heads.

Some of the theories which have been presented for their origin are that they were once part of a planet situated between Mars and Jupiter, or molten fragments of the Moon which have cooled on striking Earth. Neither of these theories is as convincing as a terrestrial origin for tektites.

Scientific analysis proves that to fuse tektites a temperature of 1,320° Celsius is required. However, it has been *proved* that when they were formed a *minimum* temperature of 2,500° C. had taken place, then they were *suddenly cooled*! (an atomic blast?)

Tektites contain the following oxides: Aluminium, beryllium, boron, calcium, iron, magnesium and potassium. Because they contain rare-earth elements the same as on Earth, some scientists *do not* believe that they came from outer space.

Two scientists at the University of Wisconsin, Drs. Larry Haskin and Mary A. Gehl, matched the rare-earth elements of tektites with Earth rocks by utilizing an exceptionally accurate method called 'neutron activation

*This surprisingly recent date was obtained by Howard Baker using the thermoluminescent dating method, accurate up to 100,000 years.

analysis'. Their discoveries *proved* that rare-earth elements in tektites and Earth rocks were almost identical.

Some tektites contain radioactive isotopes of aluminium and beryllium (Al^{26} Be^{10}) with half-lives of 10^6 and 2.6×10^6 years respectively.

Unusual tektites have been found in the Libyan Desert; analysis of these tektites discloses identical chemical composition to the Libyan Desert sand! Did the searing heat from an atomic blast, or blasts, fuse the silica content of the sand, thereby forming the glassy tektites? (which contain 75% silica; while it can also be noted that glass is made from certain types of sand).

What could be a vital Libyan Desert discovery, and which may have bearing on these tektites, is an *ancient*, highly unusual crater, situated at Lat. 22° 18′ N, Long. 25° 30′ E. Study and scientific analysis of the crater by archaeologists and scientists positively indicates that the cavity does resemble an *artificial explosion crater scar*!

At the 18th April, 1967, annual meeting of the American Geophysical Union, Dr. Bruce C. Heezen of the Lamont Geological Observatory at Colombia University said that tektites removed from the sea bed off Africa, Australia and Japan, indicate that a one-mile-diameter 'object' exploded in mid-air 700,000 years ago!

Significantly, Earth's magnetic field reversed polarity at the same time!

Some indication or proof that this 'object' may have been an atomic device, is that sharp fluctuations were detected in Earth's magnetic field on 1st August, 1958, when the US conducted an atomic test over a Pacific Ocean atoll.

It has been estimated that a 90-megatonne nuclear blast 160.9 km above Earth, could permanently change Earth's external magnetic field.

Do tektites prove past nuclear devastation and/or wars on Earth?

A NUCLEAR MISSILE WAR BETWEEN MARS AND PLANET X?

A 4,827-km-wide asteroid strip exists between Mars and Jupiter where it seems there was once a planet. Modern astronomical theories admit this probability. Early legends say that the inhabitants (of Planet X) were destroyed when a tremendous explosion disintegrated their world.

Pat Frank, author, and consultant to the US National Space Council, believes that one billion years B.C. Planet X battled with Mars – and guess who won?

Frank claimed that the existence of the mystery planet has been proved by over 200 years of astronomical research, and that it was probably the twin of Earth, with the same elements and atmosphere, but probably a little smaller. The remains of Planet X, the thousands of fragments, form what is known as the 'asteroid belt' or 'ring' between Mars and Jupiter. The largest asteroid is the 772-km-diameter Ceres.

Frank also believes that craters on Mars, the Moon, and Earth, were caused by debris (or A-missiles?) from the mystery planet, and that alternatively, Planet X inhabitants may have *accidentally* exploded their world by experiments with nuclear fission. (Or an ICBM nuclear war?)

N. J. Berrill writing in *The Atlantic Monthly*, June 1957, referred to Planet X when he said:

> 'Could such a planet have supported living beings intelligent enough, though dumb enough, to have started an atomic bomb chain reaction powerful enough to have blown their planet apart?'

A carbonaceous chondrite meteorite from the Mars/Jupiter asteroid belt, which fell in Australia near Murchison, Victoria, on 28th September, 1969, was later found to contain amino acids – the building blocks of life-form bodies.

Scientists at NASA's Ames Research Centre, Mountain View, California, who studied the meteorite, said that both left-handed and right-handed amino acids were detected. *Only* left-handed (named after their shape) amino acids are found (primarily) on Earth, so the possibility of contamination of the meteorite after contact with Earth was ruled out after this discovery, proving that these eighteen amino acids originated in space.

This find might indicate a once-existing planet in this location supporting intelligent beings.

Other meteorites from this asteroid belt strongly indicate nuclear devastation of a planet between Mars and Jupiter; some of these meteorites contain traces of uranium and thorium, two radioactive metallic oxides with vacillating mutable characteristics.

Close examination of meteorites from the asteroid belt proves that these meteoric stones once formed a planet; the crystalline structure verifies that the cooling process took place over a period of millions of years at an extreme pressure of many thousands of atmospheres. This could only take place in the interior of a planet, where the cooling process would be slow, and under very great pressure. Even more convincing is the age of these meteorites, which ostensibly approximate Earth's age, indicating that this planet may have been formed about the same time as Earth.

One interesting aspect about these meteorites is the helium content, produced partly by the decay of uranium, but *largely* by cosmic ray bombardment in space, which indicates that these fragments may have been in their broken-up form for some 300,000,000 years. One other point is that some Mars craters have been estimated to be approximately 300,000,000 years old. The 260–340,000,000-year-old Odintsovo fossil 'brain' indicates intelligent beings extant at this time!

As mentioned, some meteorites contain traces of uranium and thorium which can, and will, manifest instability, under particular trigger circumstances. Of interest in this connection is a tiny 'highly radioactive' rock brought back to Earth by *Apollo 12* in November 1969. This 4.6-billion-year-old rock contains twenty times as much uranium, thorium, and potassium as any Lunar rock, and indications are that its origin was *not* in the Ocean of Storms where it was found! Was this rock possibly related to a violent nuclear explosion somewhere in space in remote antiquity?

AN ANCIENT NUCLEAR WAR IN WEST AND NORTH SCOTLAND?

Did Atlanteans, spacemen, or others, unleash LASER-rays or atomic missiles or bombs, in an archaic war in West and North Scotland? Some years ago, unusual discoveries apparently indicated that atomic warfare once took place in Scotland. In West Scotland, an ancient fort made from boulders was found to be fused-together on top as if by sudden searing heat – LASER beams, atomic blasts – from an overhead source!

In Northern Scotland, unknown mystery symbols were found engraved on boulders. The nineteenth-century Scottish geologist Hugh Miller (1802–56) made some extraordinary discoveries in this area; he found that twelve species of animals, covering an area of 25,600 km^2, had died very suddenly and violently in remote times. The fossilized remains of fishes found in several layers, for example, were *all* twisted in violent contortions as if a tremendous upheaval had taken place. All these creatures were considered to have died in minutes, or even seconds!

What really happened here?

WHAT DESTROYED THE ARCTIC CONTINENT, AND THE MAMMOTH?

Soviet scientists have discovered new evidence for the existence of an Arctic Continent which started sinking 10,000 years ago; at this same time, a great catastrophe destroyed millions of animals in Alaska. Was there a link?

Soviet News, 17th October, 1969, states:

> 'The theory that an extensive continent – Arctis or Arctica: call it what you will – once existed in the area of what is now the Arctic Ocean is not new.
>
> 'The late Yakov Gakkel, a prominent Arctic ex-

plorer, left quite a lot of material about it. Members of the Arctic and Antarctic Institute have been following up his ideas and his papers and their findings have now been published.

'The geological structure of the American and Asian coasts of the Arctic is very much alike.

'It has long been known that they were linked by the submerged Mendeleyev and Lomonosov ridges, and the vegetation of the Taimyr Peninsula – the part of Siberia which sticks out furthest into the Arctic – is much more like that of the Canadian Arctic archipelago than that of neighbouring parts of Siberia.

'Birds of passage, even in warm countries, usually keep pretty near the coast on their migrations. Yet in the Arctic they fearlessly cross the deserted, icy wastes.

'Can this be an ancestral memory of a time – not more than 2,500 or 3,000 years ago – when the peaks of the Lomonosov Range were still above water?

'It does look as if Arctis once served as a bridge between Europe, Asia and America, over which animals and plants migrated and birds "plotted" their passage.

'Today the few islands which dot the Arctic all round the Pole are all that is left of the immense continent. But much of the Arctic is very shallow, and one can actually still trace the paths the great Siberian rivers – the Yenisei, Ob, Lena, and Kolyma – cut through the side continental shelf.'

At precisely the same time that the Arctic Continent started sinking 10,000 years ago, millions of animals died in the Tanana Valley, Alaska.

Fragmented parts of many species of animals such as the primitive elephants: the mastodon, mammoth, and other animals, were suddenly mixed with forests and plant life, the entire admixture being transported many miles away. Volcanic destruction, and devastation from glaciers, was ruled out as a possible cause. Perfectly preserved mammoths have been found in the Siberian ice with undigested food – flowering buttercups, etc. – still in their mouths and stomachs, proving that a sudden catastrophe overtook them, transported them from warmer Southern Siberia to the North, then quick-froze every cell in their bodies.

Autopsies undertaken on Siberian mammoths disclosed vital facts; skin examinations under powerful microscopes showed red blood corpuscles, which indicates a death by *water* or *gas* suffocation! A tidal wave from the sinking of a portion of the Arctic Continent, and/or plasma (hot ionized gas) from a nuclear holocaust, perhaps also causing the downfall of the Arctic Continent?

Forty million animals died in North America from this catastrophe. Equally as interesting are whale remains discovered far inland in Michigan 174.7 m above sea level, and fossil oyster shells in the Navajo Desert, although the whale remains and oyster shells may have been deposited at this time, or 2,000 years earlier when Mu and Atlantis sank.

A 1970 expedition to the Northern Yakutian shores of the Arctic, organized by the Yakut branch of the Academy of Sciences and the Leningrad Zoological Institute, discovered a huge 'cemetery' of mammoth, horse, musk ox, and sheep remains, 3 m under the frozen surface, but a still unsolved mystery was the simultaneous discovery of flints and tools belonging to Cro-Magnon Man, who was not thought to have dwelt in the frozen North. Something violent happened 10,000 years ago; what really happened?

In late 1970, Soviet geologists led by Victor Masaitis of Leningrad, discovered a giant 96½-km-diameter crater 240 km from the Arctic, in the far north of Siberia. The geologists measured the magnetic and gravitational fields and discovered that a 'mammoth explosion' had taken place. The orthodox thought is that an asteroid (from Planet X?) exploded there millions of years ago, but in view of the catastrophe there 10,000 years ago, this discovery could seem significant. The crater, 402 m deep, is filled with fragmented pieces of crushed and *once molten* rock, while giant rocks the size of 'houses' are strewn around for many kilometres. Deep fissures, gaps and breaks, cover a huge area around the crater. Possibly even more significant was the discovery of a second crater near the Yugorsky Peninsula in the Soviet Arctic. This 'conical' crater also bears witness to a tremendous explosion, with rocks strewn for a vast distance. No trace whatsoever, of any asteroid, was discovered at or near the craters.

Were these two craters in Siberia and the Arctic possibly related to the catastrophe 10,000 years ago? Did possible nuclear warfare or devastation take place?

WHAT OF THE SAHARA, NEGEV, JUDAEAN, AND SIMPSON DESERTS?

The Libyan Desert forms part of the Sahara. Scientific discoveries prove that the Sahara, and possibly the Libyan, was covered with dense jungle only a few thousand years ago. Why did this area, and others, change from heavy foliage to parched desert?

Looking at a different possibility from the accepted orthodox theories, could it be at all possible that spacemen destroyed all foliage and life in these areas with atomic weapons, thus drying up the land and soil, reducing the ground to useless sand, and irradiating the area so that nothing would grow? Remember that Libyan Desert tektites have *identical* chemical composition to the desert sand!

A 1965 discovery was four species of fossil sharks, 130,000,000 years old, at Arad in the *heart* of Israel's Negev Desert. This could indicate that the Negev Desert was once covered by salt sea which *suddenly evaporated* stranding the sharks! If the sea had receded naturally, gradually over a long time period the sharks would have left the area a long time before when the sea became too shallow, and correspondingly, the heat of the Sun too intense. Perhaps a geological upheaval cut off the sea outlet, but what could cause the geological upheaval? Natural factors, or a nuclear blast?

Several UFO researchers believe that Sodom and Gomorrah were destroyed by an atomic bomb or missile. Josephus* described it as a 'thunderbolt' (*Antiquities of the Jews*, Books I–VI).

Scholars believe that the Sodom and Gomorrah sites lie under the Dead Sea. In 1965 archaeologists discovered a 'mass grave' in this vicinity containing 20,000 skeletal bodily remains!

Also of notable interest is that the director of the UAR Atomic Energy Institute recently said that 'traces' of radioactive elements have been found at Rashi, Darniat, and the Sinai Peninsula!

In the Judaean Desert near the Dead Sea an enormous high rock depicting a boat was discovered. Could this ancient carving mean that the Judaean Desert was once covered by water which evaporated after (the?) nuclear blast which destroyed Sodom and Gomorrah?

Tektites are also found in Australian deserts. Is there a link? Australia's big Wolfe crater on the periphery of the Simpson Desert may have a more recent (meteoric?) origin than previously thought. This crater, 804 m in diameter and 30 m deep, was seen for the first time in June 1947. In 1906 'an immense ball of fire' startled residents in the area who observed it crossing the horizon, then culminating in a long rumbling vibrating explosion. This description is *almost identical* to that in Siberia two years later, when a similar 'fireball' devastated a large area. Were they (possibly?) on both occasions extra-terrestrial spaceships exploding, through faulty nuclear-powered motor systems or unstable nuclear fuel? In this connection it may be of paramount importance to note that the Tunguska blast was of equal energy to a nuclear explosion – viz: 1.1 to 2.8×10^{23} ergs, and that the surrounding air for many kilometres had a pink and green luminescence which could have been vaporized atomic matter that rose into the atmosphere and continued disintegrating.

Dr. George Baker, CSIRO (Commonwealth Scientific

*Flavius Josephus; celebrated Jewish historian, A.D. 37 to after 100.

and Industrial Research Organization) geologist, and Australian Commissioner for the International Meteoric Commission, remarked that it cannot be proven at this stage whether the Wolfe crater was caused by an impact or an explosion over the Earth's surface. Curiously, the Taiga explosion took place 5 km over the Earth's surface!

THE ATOM EATER

Did the Hindus of old possess comprehensive knowledge of atomic physics? If so, how did they see the atom without powerful electron microscopes? There are no known records of such instruments in ancient India, but still the Hindus knew of atomic physics. From whom did they learn?

Aulukya, or Kanada, 'the atom eater', a Hindu scientist of antiquity born about 850 B.C., was a most erudite fellow. Kanada, who possessed comprehensive knowledge of the atomic principles and structure of matter, was one of the most learned contributors to the Hindu philosophy of *Vaisesika*, a Sanskrit work dealing with atomic theories. Proof of Kanada's wisdom can today be found in India where reprints of his work are available; many of his theories *parallel* those of modern atomic physics!

It now seems that atomic wars described in the *Mahābhārata* are true; i.e. the 'fire-blazing discus', 'Golden arrows' (A-missiles?) 'which reduced the rocks to dust', and of wars when poison gas covered the world, and Mt Mandara was torn apart by the gods.

The *Mausola Purva* describes a missile which appeared like '10,000 suns' after spreading into three parts. A Soviet SS9 missile carries three five-megatonne atomic warheads, but it appears that the Hindus used this type of missile at least 3,000 years ago! Are those ancients today reborn as atomic scientists?

Early in 1965 Indian archaeologists discovered a completely unknown substance during excavations; it was pink and puttylike, very dense and heavy, and about the size of a grapefruit. No mention has been made as to what it may be. Is it ancient plastic explosive, or a dissipated radioactive isotope or atomic element for a Vimana or atomic weapon?

An isotope is an element with nuclei or atoms of slightly different atomic weights; example: uranium with isotopes of atomic weights 238 and 235.

King Bhojadira (A.D. 1018–1060) of the Paramar dynasty, was an intellectual, a patron of the arts, literature, and science. Near Bhopal in Madhya Pradesh King Bhoja completed an engineering feat which had no parallel in the Middle Ages. Without the assistance of modern heavy machinery he excavated a massive lake *covering an area of 640* km^2 How was this accomplished? Did he receive assistance from 'heavenly beings' with advanced technology? Did spacemen excavate the area with an atomic blast? Or did the Indians of old, with atomic knowledge, achieve the same results? This hypothesis cannot be considered as improbable; plans have been considered to utilize nuclear explosions to excavate a harbour in Australia, and a second Panama Canal.

DID ELIJAH CALL DOWN ATOMIC FIRE ON MT CARMEL?

What does the Bible *really* mean when it states that Elijah called 'fire down from heaven' (on Mt Carmel, 580m)? This 'fire' was said to have destroyed many men Refer to *II Kings 1:10, 12, 14,* 896 B.C.

Was the 'fire' atomic weapons exploding? A vital discovery seemingly proves that this is what the Bible refers to! A British scientist possesses a fragment of a 3,000-year-old *blue crystal* found on Mt. Carmel; the Mt. Carmel blue crystal is *identical* to blue crystals found in New Mexico after atomic tests!

A layer of vitrified sand turned into *green glass* was found under the sands of the Holy Land by archaeologists. Only intense heat will fuse the silica to form such glass (lightning, A-blasts, heat rays, LASER beams?).

A US development in the LASER field is a ray which

can shoot a 'drone' – pilotless plane – from the sky, and punch holes in armour plate at several hundred metres.

Some UFO researchers believe that Jericho was destroyed by extra-terrestrials utilizing sound vibrations, but this is in contradiction to actual archaeological discoveries, which prove that Jericho was devastated by *intense heat!*

Jericho was a major city located 24 km north-east of Jerusalem in the Jordan Valley. Ruins of the 180-cm-thick walls were discovered by German archaeologists near the Arab village El Riha (Jericho) early this century. The German archaeologists found evidence of Jericho's downfall by extreme heat. In June 1930 Sir Charles Marston's expedition, headed by Dr. John Garstang, archaeology professor at Liverpool University, discovered even more pertinent evidence that Jericho came to a fiery end!

Dr. Garstang said that their excavations disclosed:

> 'traces of intense fire, including reddened masses of brick, cracked stones, charred timbers and ashes . . .'

Did spacemen destroy Jericho with an A-bomb or missile, or heat rays, in 1451 B.C.?

> 'And they utterly destroyed all that was in the city . . . And they burnt the city with fire, and all that was therein . . .'
>
> (*Joshua 6:21, 24*)

Some descriptions in *Revelations* seem undoubtedly to be referring to A-bomb or missile devastation, and annihilation from germ or chemical warfare, or lethal viruses unleashed on Man. Direct reference is also made to men and creatures – not angels – who inhabit 'heaven' (*Rev. 5:3, 13*).

The following description from *Rev. 6:12, 13, 14, 15, 16* and *17 contains a prophecy for the end of the world in* what surely must mean nuclear catastrophe:

> 'And I beheld when he had opened the sixth seal and, lo, there was a great earthquake; and the sun became black as sackcloth of hair, and the moon

became as blood; And the stars of heaven fell unto the earth, even as a fig tree casteth her untimely figs when she is shaken of a mighty wind. And the heaven departed as a scroll when it is rolled together; and every mountain and island were moved out of their places. And the kings of the earth, and the great men, and the rich men, and the chief captains, and the mighty men, and every bond-man, and every free man, hid themselves in the dens and in the rocks of the mountains; And said to the mountains and rocks, Fall on us, and hide us from the face of him that sitteth on the throne, and from the wrath of the Lamb: For the great day of his wrath is come; and who shall be able to stand?'

Again in *Rev. 8:5, 7, 8, 9, 10, 11,* and *12* mention is made of similar (atomic?) devastation – and *radiation poisoning*?

'And the angel took the censer and filled it with fire of the altar, and cast it into the earth: and there were voices, and thunderings, and lightnings, and an earthquake ... The first angel sounded, and there followed hail and fire mingled with blood, and they were cast upon the earth: and the third part of trees was burnt up, and all the green grass was burnt up. And the second angel sounded, and as it were a great mountain burning with fire was cast into the sea: and the third part of the sea became blood. And the third part of the creatures which were in the sea and had life, died; and the third part of the ships were destroyed. And the third angel sounded, and there fell a great star from heaven, burning as it were a lamp, and it fell upon the third part of the rivers, and upon the fountains of waters ... *many men died of the waters because they were made bitter*. And the fourth angel sounded, and the third part of the sun was smitten, and the third part of the moon, and the third part of the stars; so as the third part of them was darkened, and the day shone not for a third part of it, and the night likewise.'

(my italics)

The 13th-century mosaic dome in St Mark's, Venice portrays 'Creation'. Two UFOs illustrated.

Above The fossilized brain of a spaceman who landed in Russia during the Carboniferous Period?

Left Fossil dinosaur footprint, and *below* fragmented bones at the Dinosaur Quarry, Colorado, where over 300 skeletons have been found. Evidence of sudden disappearance due to heat and radiation from a nuclear catastrophe?

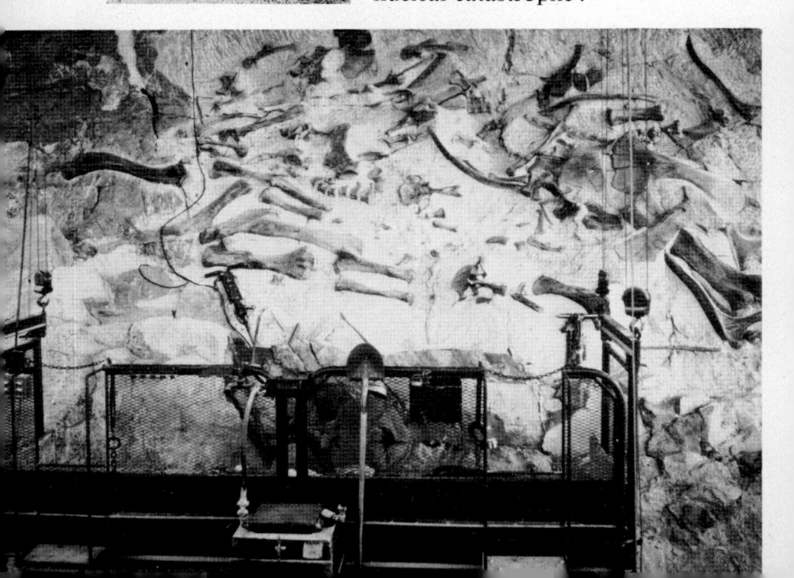

Left Ancient Egyptian scroll describing a huge 'circle of fire' that appeared in the sky, and the later arrival of many flying 'fire circles' which landed and took off again (c. 1500 BC).

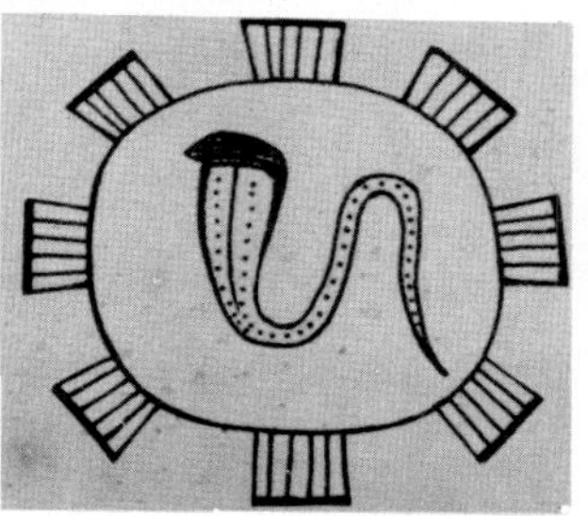

Above left and right 'Serpent' deity from ancient Egyptian papyrus (a stylized symbol for a spaceman from a planet of Barnard's Star?) and 1269 Armenian Bible illustration of Ezekial's 'flying wheel' (a UFO with four landing legs?).

Above The Great Stupa at Sanchi, India. Was it built in the shape of a UFO in the 3rd century BC?

Above left Is Christ really in a spaceship, rather than on a cloud on Mt Tábor, in this 1538 Venetian painting? Are the surrounding prophets and saints really helmeted spacemen?

Above right Madonna and Child with unusual, egg-shaped Star of Bethlehem (the spaceship in which the Christ-child was brought to Earth?). The painting, said to be by St Luke, is preserved in the Church of Our Lady of Expectation at St Thomas' Mount, Madras, India.

Below A manned spaceship shown in a 14th-century Serbian fresco.

Above left Contemporary drawing of manned rocket that appeared over Europe in 1527.

Above right Russian ink drawing of ancients worshipping spacemen as gods.

Below UFO portrayed in 1686 book *Discourses on the Plurality of Worlds*, by Bernard de Fontenelle.

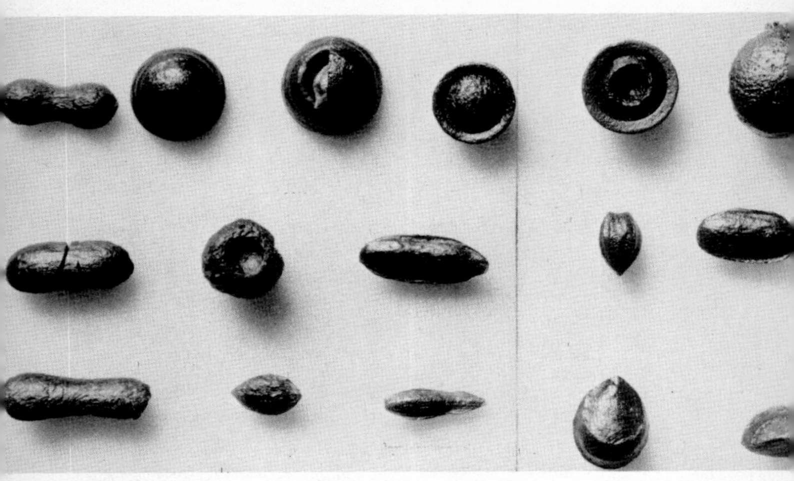

Above and left Were these Australian tektites formed by ancient atomic devastation?

Below Did the massive Ayers rock in Central Australia mark the site of the Garden of Eden? Was it damaged by nuclear missiles or LASER beams in a war waged by the serpent space beings? Was this area always desert?

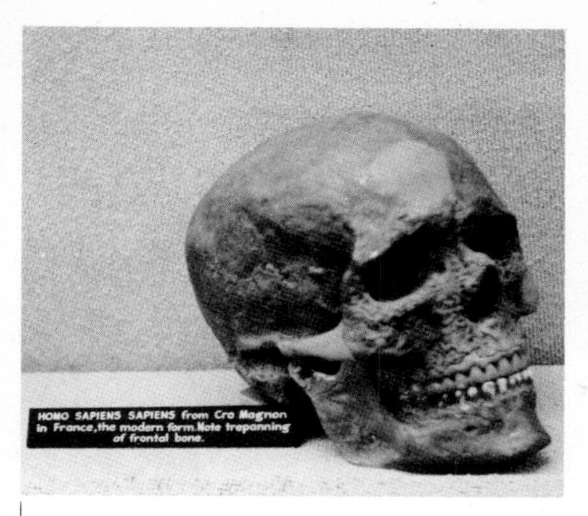

Above left Was the hole in this Cro-Magnon skull (repaired by the museum authorities) caused by brain surgery undertaken by spacemen or Atlanteans?

Above right An aboriginal rock painting, Northern Territory, Australia. Is it of a helmeted visitor in a ridged space suit?

Below UFO seen over Swan River, Perth, Australia, at 10.30 pm on January 24th, 1966, and photographed by Leslie Benedek.

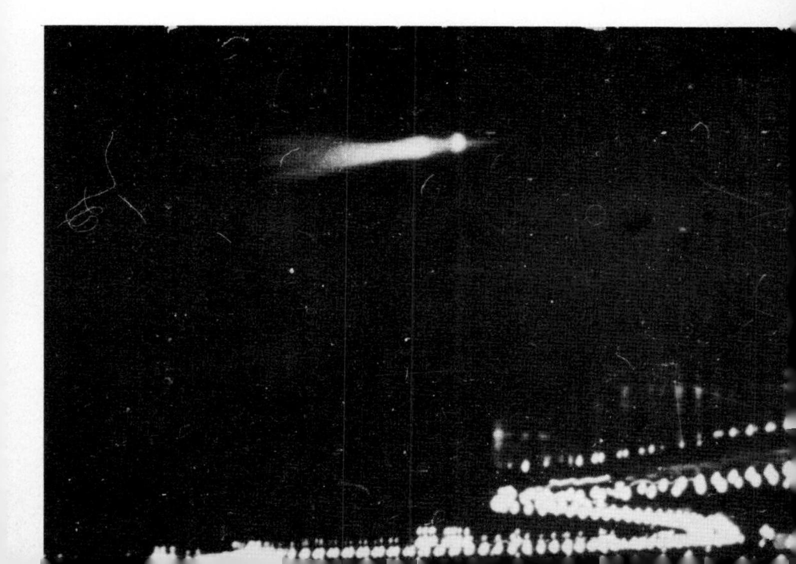

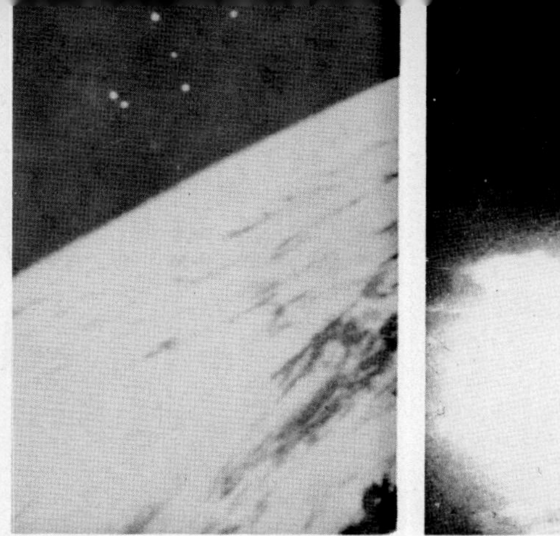

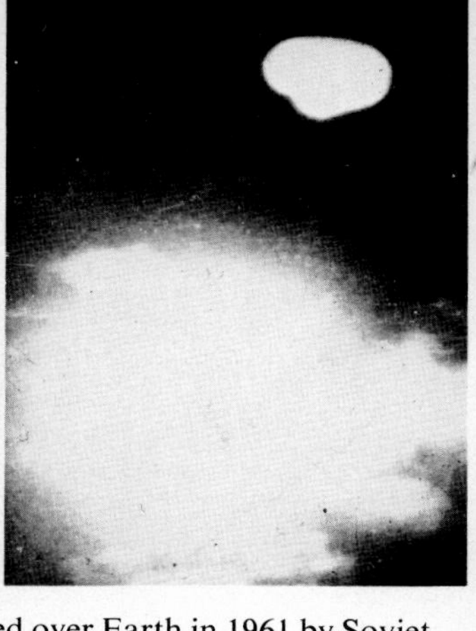

Above left Six UFOs photographed over Earth in 1961 by Soviet cosmonaut Major H. Titov.

Above right Inverted UFO photographed at 10 pm on March 17th, 1965, off the Victorian coast, by Steward Walter Jacobs, on board BHP freighter *Iron Duke*. Bright glow below is caused by the Moon.

Below 9-metre diameter UFO 'nest' photographed in N. Queensland by P. V. Vignale. The 23 cm. reeds were pressed down in a clockwise direction, and radiation tests revealed only Beta counts in one reed sample. A UFO was seen to rise from 'nest' by banana grower George A. Pedley, at 9 am, January 9th, 1966.

Why would 'many men' die 'of the waters because they were made bitter'? Does this refer to radioactive contamination from atomic 'fallout'?

Rev. 9:1, and *2* describes another 'star' (or A-bomb?) which fell 'from Heaven':

> 'And the fifth angel sounded, and I saw a star fall from heaven unto the earth ... And he opened the bottomless pit; and there arose a smoke out of the pit, as the smoke of a great furnace; and the sun and the air were darkened by the reason of the smoke of the pit.'

Rev. 16:8, 9, 10, 12, and *18* outlines what seems to mean atomic devastation:

> 'And the fourth angel poured out his vial upon the sun; and power was given unto him to scorch men with fire, And men were scorched with great heat ... And the fifth angel poured out his vial upon the seat of the beast; and his kingdom was full of darkness; and they gnawed their tongues for pain ... And the sixth angel poured out his vial upon the great river Euphrates; and the water thereof was dried up ... And there were voices, and thunders, and lightnings; and there was a great earthquake, such as was not seen since men were upon the earth, so mighty an earthquake, *and* so great.'

Some references in *II Kings* have a familiar ring when comparing them to nuclear warfare or devastation descriptions! For example:

> 'Behold, thou hast heard what the kings of Assyria have done to all lands, by destroying them utterly ...'
>
> (*II Kings 19:11*) 710 B.C.

> '... now I have brought it to pass, that thou shouldest be to lay waste fenced cities into ruinous heaps.'
>
> (*II Kings 19:25.*)

> 'And it came to pass that night, that the angel of the Lord went out, and smote in the camp of the

Assyrians an hundred four-score and five thousand: and when they arose early in the morning, behold they were all dead corpses.'

(*II Kings 19:35.*)

What did Joel mean in 800 B.C.? He said:

> 'The earth shall quake before them; the heavens shall tremble: the sun and the moon shall be dark, and the stars shall withdraw their shining.'
>
> (ibid. *2:10*)

Was Joel prophesying, or giving advance warning to future generations about the horrors of nuclear war?

Jesus said the same over 800 years later:

> 'Immediately after the tribulations of those days shall the sun be darkened, and the moon shall not give her light, and the stars shall fall from heaven, and the powers of the heavens shall be shaken.'
>
> (*Matthew 24:29.*)

What did they *really* mean?

> '. . . the heavens shall pass away with a great noise, and the elements shall melt with fervent heat, the earth also and the works that are therein shall be burned up.'
>
> (*II Peter 3:10.*)

With the blue crystal discovery on Mt. Carmel, it seems that the ancient of Biblical days *really did* know of nuclear warfare!

Were Joel and Jesus warning future generations?

NOSTRADAMUS AND EDGAR CAYCE MAY HAVE PREDICTED A-WARFARE

Nostradamus (1503–66), or Michel de Notre Dame, a French astrologer with uncanny powers to foretell future events, published his book of 'Oracles' or prognostications in 1555.

His works have been of great interest since, as so many of his prophecies eventuated this century (including the A-bombing of Nagasaki), which proves that he was a man of rare powers unexplainable to science.

Nostradamus enigmatically seemed to infer that Earthmen almost start, or start, an atomic war in 1999, but are stopped by spacemen.

In *9:44* he states:

> '. . . The contrary of the positive ray shall exterminate all (negative ray – nuclear radiation?)
>
> 'Before it happens, the Heavens shall show signs.'

According to Nostradamus this could happen between 23rd November and 21st December, 1999!

Still another reference to 1999 seems to mean that in July of that year, before the impending possible catastrophe, a spaceman from another planet arrives on Earth and attempts to persuade the world's leaders to avert nuclear holocaust:

> 'The year 1999, seventh month,
> A great king of terror will descend from the skies,
> To resuscitate the great king of Angolmois,
> Around this time Mars will reign for the good cause.'
>
> (*10:72.*)

The stanzas also seem to contain a reference relating to scientists who warn mankind about the horrors of nuclear war, but who are spurned:

> 'The most learned in the celestial sciences,
> Shall be found fault with by ignorant princes,
> Punished by a proclamation, chased away as wicked,
> And put to death where they shall be found.'
>
> (*4:18.*)

According to Nostradamus, after we have overcome this dark period, a new Renaissance will follow, which corresponds with the Biblical prophecy of a Utopian age soon to come.

> 'And he shall judge among the nations, and shall rebuke many people: and they shall beat their swords into plowshares, and their spears into pruninghooks: nation shall not lift up sword against nation, neither shall they learn war any more.'
>
> (*Isaiah, 2:4.*)

A most unusual observation can be noted in the parallel of Edgar Cayce who also predicted a possible world catastrophe between 1998 and 2000. Cayce was *equally* as accurate as Nostradamus in prophesying future events. Was Edgar Cayce, Nostradamus reincarnated? Cayce believed that he had once lived a past life as an Egyptian priest, and another as a Persian physician.

Still another notable point in this context was that when Galileo's life came to a close in Italy, Newton was born in England that same year (1642) and continued the work where Galileo left off. Galileo discovered how things moved, but Newton discovered why, and advanced/refined Galileo's theories. Was Galileo reborn as Newton?

ATOMIC BLASTS ON MARS?

On 9th December, 1949, the distinguished Japanese astronomer Tsuneo Saheki witnessed through his telescope, what appeared to be a tremendously powerful explosion on Mars. Saheki, who had studied Mars since 1933 in the category of a specialist, reported that a 'brilliant glow' lasting several minutes followed the bewildering explosion; then a 'luminous yellow-greyish cloud' appeared. The dimensions of the 'cloud' were estimated at a 1,120-km-diameter, with a height of 64 km 360 m.

Saheki said:

> 'It was undoubtedly an artificial atomic explosion ... even more powerful than the H-bomb. That could only have been set off by highly advanced beings ...'

On four occasions between 1937 and 1954, astronomers

at Osaka Observatory watched 'brilliant' light flashes on the Martian surface. Volcanic eruptions were *ruled out*, as the light flashes were far more luminous and intense than any volcanic activity! On the first and second 'explosions' the 'brilliant' 'light' was observed for 300 seconds (each), and 5 seconds each for the remaining two 'explosions'.

Mariner photos of 'recent' Mars craters might be interesting in this respect.

In his thesis *A Visitor From Outer Space*, Aleksandr Kazantsev said:

> 'After the great opposition of Earth and Mars in 1956, A. A. Mikhailov, Director of the Pulkovo Observatory and Corresponding Member of the USSR Academy of Sciences, reported to a gathering of scientists at the Leningrad Scientists Club in Lesnoye, that the Pulkovo Observatory had registered an explosion of tremendous force on Mars. Judging by the fact that the consequences of the explosion had actually been observed through telescopes . . . the explosion had to be attributed to nuclear explosion rather than to anything else. It is difficult to imagine a nuclear explosion on Mars that was not deliberately caused. It is very likely that the explosion was made for some constructional purpose. Thus, the Pulkovo Observatory observations can serve as one of the proofs in favour of the existence of rational life on Mars.'

Kazantsev is a journalist, an adventure, science, and science-fiction writer, and an engineer by profession.

RADIOACTIVE VAULTS AT MITTELFRANKEN

In 1961 an extraordinary discovery was made at Wasserburg Somersdorf castle in Germany's Mittelfranken Province. In the castle, the ancestral home of Baron von Crailsheim, several family vaults more than 250 years old were opened. Within the vaults were remarkably preserved bodies of men, women and children still with hair and fingernails!

Leading European and US scientists discovered that the bodies and vaults were *strongly radioactive*; this factor had obviously preserved the bodies! Radioactivity was also detected in the castle lake.

The scientists could offer no explanation for the radioactivity which was still giving 'strong' gamma count readings even after hundreds of years.

There is no natural uranium or other radioactive deposit under the castle; the source of the radiation remains unknown.

A NUCLEAR SHELTER IN THE TRANSVAAL?

In 1964 a complete underground village was discovered in the Kleinfontein Valley in the Transvaal, South Africa. Archaeologists who located it were struck by its *similarity* to modern nuclear war shelters, but this underground village was built in antiquity!

The village, in caves deep underground, is comprised of huts with mud walls about 30 cm thick; each cave contains approximately twenty huts. At ground level are the remains of another village constructed by the same race. It is thought that the mystery people – of whom there are no records – took refuge in the underground village in times of (nuclear?) war.

North and West Africa are tektite locations; did nuclear warfare take place in ancient Africa?

It seems there are both 'sons of light' and 'evil angels' (Dead Sea Scrolls/Apocryphal works), as we have both on our own planet.

DELIBERATE CONTROL IMPLANTS?

Many or most children, while still in infancy, unfold in their dream sequences, most unusual recurring 'nightmares' of a terrifying nature, which interestingly follow a strangely similar pattern. Three examples of such dreams, are: falling endlessly through space without contacting solid ground (as if dropped from a spaceship?); being whirled; and attempting to run – to escape – from something frightening or horrifying, which is closely approaching, only to find one's legs becoming leaden; and/or losing motive power, and being able to move only very, very slowly.

Several other similar-type dreams could be classified in this same apparent pattern. The enigma is why so many children the world over experience these dreams. This could seem a curious theory to advance, but remembering the words of Maxim Gorky: 'all that seems mysterious has a very definite basis in reality ...' – and looking beyond the Freudian and/or present-life psyche, could it be that possibly millions of years ago on other planets, or Earth, many people were brainwashed and controlled into abject submissiveness, by evil spacemen, and/or Earthborn space beings, using electronic and associated methods which would create the effects in the dreams described above?

Are these dreams racial memories or past-life *experiences* recorded in the memory banks of Man's subconscious or computer (brain)? At a certain period of life, while quite young, many children may be very sensitive to receiving certain past-life memories which may impinge on the subconscious?

Does Man *still* react unknowingly as a child and adult, to these possible archaic fear/control implants?

6. DID EARLY NEW ZEALAND MAORIS LEARN ASTRONOMY FROM SPACEMEN?

'Home of the gods'

> (Translation of the Maori name Pahiatua, a borough in New Zealand's North Island.)

In this chapter, the submergence of Mu is again examined (refer to Chapter 1). Could possible reasons for sinking of the Arctic Continent, Atlantis, and Mu, and ancient nuclear holocaust mysteries outlined in Chapter 5, possibly be in any way related? Or did Mu sink due to natural geological causes?

From where did the first Maori settlers come? The Maoris are believed to have migrated to New Zealand – 'Aotearoa' – from the Hawaiian islands, although ethnologically, similarities exist between the Maoris, Malay, and Indonesian races, and certain races in both North and South America. Other suggested origins even include India, Ur, Egypt, and Mongolia; it is uncertain where the Maoris *originally* dwelt.

An anomaly associated with the Maori 'migration' legends is difficult to identify with observed star navigation facts. If as thought, they travelled south-west, and used the stars which rise in the east as navigation fixes, or kept the prows of their canoes 'to the left of the Sun or Venus' as recorded in another account, then it seems likely that the landing site would be South America, but, strangely, the gods (spacemen?) are given some credit for having guided the ancestral vessels to New Zealand.

The mystery is further compounded by the discovery that the chief deity who guided the ancestral vessels from 'Hawaiki', the legendary homeland, to New Zealand, was 'Uenuku' or 'rainbow'. For many years Maoris searched for a sacred wood carving of Uenuku, believed to have arrived with the first canoes; the search was quite unsuccessful, until accidentally the sacred carving was discovered in 1905 in Lake Ngaroto, 17.7 km south of Hamilton.

This remarkable carving is not the original, but was carved from New Zealand Totara wood, as a copy, when the Maoris arrived.

The stylized carving of Uenuku, 600 to 700 years old, stands 2 m 63 cm tall with an unusual 'coxcomb' and a row of truncated chevrons – a zigzag outline cut-off at the tips – running down the chest.

Uenuku, the most sacred Maori relic in existence, is credited with magic powers; it is believed to have 'an intensely comforting or disturbing effect on people'. Is it possible that a 'condensed dynamism' has been instilled into the carving as a protective agent, possibly similar to the unidentified protecting power in the Egyptian pyramids?

In the entire pantheon of Maori deities there is no carving similar to Uenuku; it is unique. Maori legend relates that Uenuku lived in the hole in the head of the carving; by visualizing a man's head inside the opening, it resembles a stylized space helmet with emanating rays (the 'coxcomb').

Was Uenuku a spaceman? In all of Polynesia there are only two carvings similar to Uenuku; these are in the Bishop Museum, Honolulu. The most interesting of the two carvings portrays a figure with a curved loop over his head and attached to the shoulder; this figure is also crowned by a coxcomb similar to that on Uenuku. A very similar carved bone pin, belonging to the Chalcolithic Period was found in Beer Sheba, while, amazingly, *identical* carvngs to Uenuku are kept in the Sikh temples in Northern India!

Who was Uenuku?

Discoveries made by Soviet ethnologist Julia Sorokina (Chapter 1), relating to similar blood-grouping, and *almost identical* legends between the Chukchi and Eskimo with the Polynesians, may be more than incidental in view of the Maori 'migration', especially since it was discovered – from legends – that the Eskimos were transported from

Mongolia, among other places, to the North by massive spaceships. Mongolia is one suggested original homeland for the Maoris.

Further confirmation of a link between Maoris and Eskimos, is the North Alaskan Eskimo legend of the omnipotent 'Tulugaak', or 'Raven'. The deity Tulugaak was believed to have brought up land from the sea, caused differentiation between day and night, and to have possibly created Man. This legend is very closely paralleled by the Maori deity Maui, who was believed to have 'fished' New Zealand from the sea, 'snared the Sun', and to possess magical powers.

When the Maoris arrived in New Zealand they found the Morioris, a darker-skinned race, who finally became extinct in 1934. Not a great deal is known of Moriori history, while their first point of origin remains uncertain.

The Morioris were not Maoris as has been suggested. Comparison of anthropological cranial/skull measurements between Maoris and Morioris, the average figures calculated from more than 50 sets of cranial/skull measurements taken with each group, show subtle but distinct variations in length, breadth, Cephalic Index, Zygomatic width and Index, and Nasal height and width.

Many Morioris had a decidedly Semitic cast to their noses, which recalls to mind possible remnants of a remote link with 'the ten lost tribes' of Biblical times – tribes comprised of Assyrians and Israelites.

An interesting 'blood relationship' and physical appearance exists between Maoris and Haida Indians of the Queen Charlotte Islands, British Columbia, and Alaska, while their wood carvings are similar, but for those interested in a South American link for the Maoris, new discoveries revealed the still extant Yuri, an ancient Colombian Indian race previously believed to have been extinct at least 50 years.

Recently, surviving Yuris were located in south-east Colombia between the Caqueta and Putumayo rivers at the Puré headwaters. The Yuri speak a language unknown to other South American Indians; their features resemble Maoris, and they practise face-tattooing like the ancient Maoris. Like the ancient Maoris, they also possess a similar type of stone axe.

If the Yuri are a possible ancestral link with the Maoris, then they would not have migrated from Colombia under their own volition. Since time immemorial they have stayed in their confined area, unknown to most other Indian tribes in South America. If there is a possible remote link between the Yuri and Maoris – which cannot be ruled out, if not in recent centuries then perhaps even further back, on Mu? – could it be remotely possible that the Yuri may have been taken from Colombia by other beings (in what means of conveyance?) and resettled in New Zealand?

Summarizing all evidence at the end of this chapter, a definite theory for the Maori migration is advanced. At this stage we are looking at several different ideas and possibilities, some of which may be considered quite unorthodox.

At Poukawa, 17.7 km from Hastings, Mr. T. R. Price, an amateur archaeologist, discovered evidence of human habitation dating back before the Taupo eruption of A.D. 135. This has now set the date of human habitation in New Zealand back by *at least* several hundred years! Further research disclosed burnt bone carbon-dated at 3,180 years of age, and ash dated at 3,370 years. Underneath the ash Price discovered traces of *post holes* in a 21-m straight line. This *proves* that Man settled in New Zealand in the vicinity of 4,000 years ago. As expected, there is orthodox opposition to these finds.

Who were those unidentified beings living in New Zealand thousands of years before discovery by the first Europeans? Ethnological studies indicate that Polynesians could not have settled in New Zealand so far back, but Maori legends from later periods indicate that unusual alien beings were either resident, or visitors in New Zealand.

When the Maoris arrived, apart from the Moriori people there were three types of extremely unusual beings in New Zealand. The first was Maeroero, a savage entity covered in long hair who resided in the mountains;

the second was a 'half-human' creature called Aitanga-a-nuku-mai-tore who dwelt in tree houses, and the third, undoubtedly the most fascinating of all, were the Turehu or Patupaiarehe 'fairy people'.

The fairy people, who clothed themselves in white raiment, were short-statured with fair skin, blue eyes and flowing reddish-coloured hair. Maori legends say the fairy people resembled human beings, but possessed strange magical powers. Upon arrival of the Maoris they fled to the forests and the summits of mountains, particularly Mt. Ngongotaha, west of Rotorua.

Were the fairy people space beings? Pointedly, they were also known to the Maoris as 'atua ririki' (little gods) or 'iwi atua' (tribe of gods). The fairy people may have had a link with certain New Guinea natives, Central Australian aborigines, and Maoris in Nelson, N.Z., all with reddish hair.

An old Maori of the Awa tribe of Te Teko once said: 'There is no limit to the world according to Maori belief, and I was taught that there are persons in the heavens ... other supernatural offspring remain on Earth.'

Who were, or are, the 'supernatural offspring'? The children of a union between space beings and Earthlings, or were the 'offspring' other space beings? Of interest in this connection, were the lighter-complexioned Maoris called 'Urukehu', or 'fair hair', whose ancestors were 'Whanau-a-Rangi', which means 'offspring of heaven'.

Were the Urukehu the children of an ancient union between space beings and Maoris?

The first part of the name Urukehu, is 'Uru', as is the last part of the aborigine name for Ayers Rock. Enigmatically, one Maori legend refers to the ancestral home as 'uru'. This name is a puzzle to ethnologists attempting to trace the origin or homeland of Uru, but an archaic Maori definition of Uru is very interesting. The definition is 'Uru the Red' or 'Gleaming One'; does this refer to Mars 'the Red Planet'?

In Chapter 11 we will find that a little-known tribe in a remote area of the world, call themselves 'Uru'!

KAITANGATA MARRIED A GIRL FROM ANOTHER PLANET

An indication of marriage between Earthlings and space entities producing children, is in the following condensed Maori legend:

The tale is of a beauteous goddess who dwelt in the heavens; her name (in Maori) was Waitiri. The divinity Waitiri who was in love with a mortal, a noble Maori warrior named Kaitangata, descended to Earth one day and married her devoted warrior. After remaining in Aotearoa for several years, Waitiri bore her husband two sons and a daughter, but due to strife between Kaitangata and his space wife, the goddess decided to return to her home in the sky. Subsequently, a 'cloud' (UFO?) descended from the sky and took Waitiri back to her original home among the stars. The Maoris named a town 'Kaitangata'; was this the site where Kaitangata and Waitiri lived?

Not only New Zealand, but throughout Polynesia, Micronesia, and Melanesia, are ancient legends of gods and goddesses descending from the skies in antiquity, and even of Earthlings being taken on trips to the Moon.

A story similar to that of Waitiri and Kaitangata is the Chinese legend of a cowherd abducting one of seven space 'fairies' as they bathed in a stream; the cowherd married the space fairy who bore him a son and daughter, but one day, the 'fairy' managed to escape and ascend to 'Heaven' on a 'cloud' (UFO?)

An almost identical parallel to the Chinese legend can be found in Melanesian legends from west and north New Guinea, New Caledonia, and New Hebrides. Who *really* abducted the space lady?

THE KUMARA WAS BROUGHT FROM LYRA, OR VENUS

A variant of sweet potato named 'Kumara' grows in New Zealand; of all the Maori foods, it is the most favoured, and was believed to have originated in the heavens.

Maori belief is that the Kumara is not indigenous to Earth, and at one time was not to be found anywhere on our planet. Maori legend records that the Kumara tubers lived in the sky and were protected by the star 'Whanui' (Vega). At an opportune time Rongo-maui flew to the sky, stole some Kumara tubers from Whanui and brought them back to Earth. Does this legend, in reality, refer to the gift of Kumara tubers to the Maoris by beings from a planet orbiting around Vega, or from Vega's constellation Lyra?

A 4,000- to 6,000-year-old Sanskrit word may be explicable in this context; the word, probably derived from the Vedic hymns, is 'Sanat-Kumāra – which was, according to ancient records, the name of a race of Venusians. There does seem to be an ancient link somewhere!

The original sacred carved stone figurine of the Maoris' Kumara god which was presented to Captain Cook, and is now kept in the 'Auckland Institute and Museum', has curious features. Its grotesque appearance is deliberate, as with other carvings of Maori gods; the reason is that it was forbidden to carve in the exact likeness of the gods, presumably to disguise their real appearance and so keep secret their visitations to Earth!

The carving has three fingers on each hand, and two toes on each foot; does this really represent mechanical hands and feet as part of a space suit? Japanese experts are puzzled by a figurine from the Jomón period (7000 to 200 B.C.), which, like the Kumara god, has only three fingers on each hand!

Another item believed by the Maoris to have had its origin in the heavens, is the 'Maire' shrub.

PERHAPS SPACEMEN TAUGHT ASTRONOMY TO THE MAORIS OF OLD?

The Maoris named the stars 'little suns' or 'Ra Ririki' hundreds of years before the Europeans conceived of them as such! And, according to Elsdon Best (1856–1931), New Zealand's most knowledgeable authority in the study of Maori ethnology, the ancient

Maoris knew that Earth was round at a time when Europeans believed it to be flat!

Maori men of the 'Tohunga' (adepts), who supposedly knew the secret of levitation, and who knew 300 stars by name, possessed stellar knowledge in advance of European astronomical knowledge of the day!

If it is thought that these statements can be explained away, then the following facts offer even more convincing proof of advanced astronomical knowledge among the ancient Maoris.

The *only* astronomical devices possessed by the ancient Maoris were sticks placed in the ground, whereby the movements of certain stars could be observed by lining them up with the sticks; yet they knew of an excess of seven stars in the Pleiades, which, incidentally, were venerated by the Maoris of old.

Persons with normal eyesight can detect six stars in the Pleiades; exceptionally sharp eyesight may detect a seventh, but the Maoris of antiquity knew of 'several' stars in excess of seven in the Pleiades! *Only* binoculars or a telescope will show this many!

Some UFO writers believe there are 12 planets in our solar system; others are too far away for our telescopes to detect. Soviet scientists have calculated that our solar system could be considerably larger in diameter than is known at present, and that there could be several other planets beyond Pluto. Further planets would, of course, be extremely difficult to find, due to the enormous distance of Pluto from the Sun.

Our telescopes cannot at present discover additional planets beyond Pluto, but the ancient Maoris of the Takitumu tribes adamantly believed in the 'twelve heavens', or 'Nga-Rangi-Tuhaha'.

Beings called 'Whatukura', or messengers of God, were believed to reside on the twelfth heaven.

The ancients in the Near, Middle, and Far East knew of only 'seven heavens', namely the Sun, Moon, Mercury, Venus, Mars, Jupiter, and Saturn, in reality only five planets, or six including Earth. Shortly we will discover that the separate 'heavens' of the ancients may, in reality, have been heavenly bodies, although several gradations or regions of the Astral Heaven or World possibly do exist in the ethereal plane.

Will astronomers one day discover three additional planets beyond Pluto in our solar system?

Other Maori tribes spoke of ten heavens, and still others of twenty. Did their deities or ancestors come from solar systems with planets of these numbers?

Strangely, some of the ancient Maori names for heavenly bodies were identical to those used in Egypt and Babylonia of old; i.e. the Maori name for the Sun was 'Ra', as it was in Egypt and Babylonia. Still another name for the Sun in Egypt was 'Kau'; the Maoris used this name in connection with movements of the heavenly bodies!

An archaic link also existed between the Maoris and India; the Maoris possessed three 'Tapu' or consecrated 'baskets of knowledge' received from the Supreme Deity, (actually from the twelfth heaven), while the Hindus of antiquity possessed three sacred books of knowledge!

In days of yore the people of India spoke of the 'Apas' or 'cloud maidens'; the Maoris referred to 'Hine-kapua' the 'cloud maid'. *Who* were the cloud maidens?

The most startling observation of all is that the Maoris of antiquity possessed knowledge of the bands of Jupiter and the rings of Saturn! These things *cannot* be seen with the unaided eye, but *only* with a telescope!

Guy Murchie says in his book *Music of the Spheres* that:

> 'One of the great mysteries connected with Saturn, is the still unanswered question of how the ancient Maoris of New Zealand knew about her rings – for there is evidence that they did have a Saturn ring legend long before the days of Galileo. Could they have had concave parabolic mirrors in some long-forgotten civilization? Is it conceivable that they descended from a great "lost continent of Mu" in the Pacific Ocean that had advanced to the discovery of optical lenses before vanishing practically without trace?'

A large land mass sank in the Pacific thousands of years before the Maoris arrived in New Zealand, but we will look at this in a moment. Although optical lenses of extreme age were excavated in Australia, there is absolutely no evidence that the Maoris of old possessed concave parabolic mirrors or optical instruments, only sticks placed in the ground as astronomical aids. A telescope would show Jupiter's bands and Saturn's rings, but even a telescope would be most unlikely to reveal additional planets beyond Pluto, of which indications are that the Maoris of the Takitumu tribes did know about!

Did the Maoris learn astronomy from advanced beings? Due to the obvious similarity between Maori astronomical knowledge and ancient Eastern astronomical knowledge, the reader will probably conclude that an ancient link existed in this context. This could indicate a possible point of origin for the Maoris.

In this chapter, diverse ideas have been forwarded for the origin of the Maoris, and how they arrived in New Zealand, but the following is what I consider most likely.

Thousands of years ago, the Maoris lived on Mu in the Pacific; archaic legends state that highly advanced civilizations dwelt on land masses in the Pacific, and even preceded Atlantis in advancement. In fact, they were reputed to be the very first sites where Man made his advent upon Earth!

The Maori men of the Tohunga, among several or many other peoples, either brought astronomical knowledge with them from home planets when they were first brought to Earth in spaceships and resettled on Mu, or all these peoples learned astronomy on Mu from advanced beings. But who were the advanced beings? Spacemen, or Earthlings from other parts of the globe? What of the astronomical knowledge of ancient Egypt, Babylonia, and India, so similar to Maori astronomical knowledge? Did the Egyptians, Babylonians, and Indians travel to Mu to teach the inhabitants astronomical knowledge (obtained originally from where?)? This idea has a chief weakness; identical carvings to

Uenuku are kept in the Sikh temples in Northern India which seems to indicate a link with Mu of the *same* time period, not earlier. So the peoples of India, for example, could not have travelled from India to teach the inhabitants, for they too must have come from Mu! This hypothesis is supported by old legends referring to early settlers in India who originally dwelt on a land mass in the Pacific!

Louis Jacolliot's book *Histoire des Vierges: Les Peuples et Les Continents Disparus*, refers to the earlier Lemuria preceding Mu, when it is stated that:

> 'One of the most ancient legends of India, preserved in the temples by oral and written tradition, relates that several hundred thousand years ago there existed in the Pacific Ocean an immense continent which was destroyed by geological upheaval, and the fragments of which must be sought in Madagascar, Ceylon, Sumatra, Java, Borneo and the principal islands of Polynesia.'

The Maoris apparently *never* possessed knowledge to grind concave parabolic mirrors or optical lenses to mathematical formulae, so it must be assumed that their advanced astronomical knowledge of certain heavenly bodies – astronomical knowledge which *cannot* be obtained with the unaided eye – was handed down through many generations from their forefathers on Mu, who either brought astronomical knowledge with them from other planets, or were taught on Mu by space-being teachers!

Ancient optical lenses have been excavated or discovered in Australia, Central America, China, Ecuador, Egypt, and ancient Nineveh on the Tigris in old Assyria; but even if the Maori forefathers on Mu possessed telescopes – which in itself would be a remarkable discovery – it still would not explain the mystery of the twelve heavens – Nga-Rangi-Tuhaha – 'the bespaced heavens' known to the Takitumu Maori tribes of the East Coast. If the twelve heavens were three additional planets beyond Pluto, in our solar system, then a telescope probably would not show them! It is strongly suspected that additional planets exist beyond the ninth planet Pluto, but less than one-eighth of Pluto's orbit is known, and as Pluto takes almost 250 years to complete one orbit of our sun, any deflection from Pluto's orbit, by the force of gravity exerted by a tenth planet, would be extremely difficult to calculate at present, and with our current telescopes we still cannot see (three) additional planets beyond Pluto!

Scholars believe that the 'seven heavens' of the ancients were seven 'heavenly regions', i.e. the lowest region is the region of the stars, while the highest is the abode of God; this belief originates from the Cabbala which represents the later Rabbinic conceptions. However, during research I have found sufficient evidence to indicate that the 'seven heavens' may have been heavenly bodies in our solar system, again namely: the Sun, Moon, Mercury, Venus, Mars, Jupiter and Saturn. If the Nga-Rangi-Tuhaha of the Maoris was in fact our solar system, even if the Sun and Moon were counted, it would still give two additional planets beyond Pluto, or one if Earth was counted. It is also unlikely that the twelve heavens of the Maoris were signs of the Zodiac, which they knew as 'ara matua' ('the main road'). Other factors which indicate that the twelve heavens were a solar system, possibly ours, is that the beings known as 'Whatukura' or messengers of God were believed to dwell on the twelfth heaven; which was also the place from which the three sacred occult baskets of knowledge were obtained! The ten and twenty heavens believed in by other Maori tribes indicate other solar systems, not the Zodiac!

The Takitumu men of old regarded as true a subject that I myself have reflected upon. They believed that the heavenly bodies are arranged in their precise courses by the gods. This would mean that solar systems are not entirely the work of the Supreme Creator, but that advanced spacemen *also* contribute with their unbelievably advanced sciences, to space and regulate the orbits of planets in a new solar system! Perhaps with rays/pulses?

Assuming that highly advanced space beings live on other planets, then the probability exists that galactic teachers lived on Lemuria or Mu; but who were the rulers, and who was Uenuku? Whoever he was, Uenuku was greatly esteemed in antiquity; identical carvings to Uenuku are preserved in the Sikh temples in Northern India, which seems to indicate a *definite* link with Mu, a similar carved bone pin resembling Uenuku was found in Beer Sheba, and similar carvings to Uenuku are kept in the Bishop Museum, Honolulu. Both the Hawaiian carvings most definitely represent man-like beings!

It could be that Uenuku was the chief deity – or spaceman – on Mu, which up till 12,000 years ago probably included the Hawaiian islands when they formed part of a huge land mass. So, it was probably from the Hawaiian islands that the Maoris left on their migration to New Zealand, bringing with them in the 'Tainui' canoe, a carving of the greatly revered deity they call Uenuku. But, several deities (spacemen?) are credited with guiding the canoes to Aotearoa; one was Uenuku. Or perhaps it was the personified carving of Uenuku, credited with magical powers?

A Madras Indian, who in 1966 visited New Zealand's Te Awamutu Gavin Gifford Memorial Museum where Uenuku is kept, was amazed when he saw Uenuku; he called it 'God'. Pertinent evidence from ancient lexicons, chronicles, legends, and codex fragments indicates that 'God', to many ancients, was an exceedingly advanced spaceman ruler and/or teacher, or even rulers and teachers. It seems Uenuku was the most revered and esteemed spaceman ruler/teacher on Mu, and/or who was either, or also, the commander of a spacecraft fleet which brought the Maoris – among other races, or racial types – to Mu.

Of the two Pacific continents, Lemuria and Mu, it is Mu which is probably the most likely to have been the Maori homeland, due to the more recent event of the Maori migration, and the fact that Mu, according to Col. James Churchward, sank only a few thousand years ago; whereas Lemuria, or part thereof, probably sank, or started sinking, long before this time. Col. Churchward does not report yet another large land mass in the Pacific sinking after Mu.

Maori legend seems to indicate that Uenuku was a spaceman who dwelt on Earth then returned to the sky. The legend recounts that Uenuku once lived on Earth, then travelled to the sky where he became the personified 'rainbow'! Perhaps Uenuku represents not just one, but several spacemen rulers/teachers on Mu?

Referring again to Aotearoa; the rising of certain stars heralded planting time for the Maoris of old. One legend from probably a Ngapuhi tribe was related by a learned Maori, who said:

> 'I have spoken of these stars as a token of regard for the beings who directed our ancestors and elders, now lost to this world.'

Who were those 'beings' who directed the Maoris' 'ancestors and elders'? Spacemen?

A legend from the Awa people refers to a 'supernatural' being, or Atua, who descended at Te-Hapua o Rongomai near Island Bay, Wellington, in days long since gone.

Finally, the ancient Maoris believed that the secret of making fire was a gift brought by the deity 'Auahi-turoa'; was he the Greek deity Prometheus who brought the gift of fire? Both the Maori and Greek fire origin legends are very similar!

7. WHY DID MU AND ATLANTIS SINK 12,000 YEARS AGO?

'Time is like a river made up of the events which happen, and its current is strong; no sooner does anything appear than it is swept away, and another comes in its place, and will be swept away too.'

(Marcus Aurelius Antoninus, A.D. 121–180.)

Why, if Mu was the original Maori homeland, did it sink? In Hans Bellamy's *Great Idol of Tiahuanaco* Hoerbiger, the Austrian engineer and mathematician, expounded his hypothesis that our Moon was 'captured' by terrestrial gravitational attraction some 12,000 years ago. According to Hoerbiger the resulting stresses on Earth from the arrival of the Moon caused the sinking of Atlantis and the Biblical Flood. One may also note that Col. James Churchward claimed that Mu sank 12,000 years ago.

One old Tibetan chronicle refers to antediluvian times when Earth did not possess a moon. This does not mean that the Moon did appear at or after the Flood, it may have appeared before.

Legends of a once moonless Earth are also paralleled in the Slavic fairy-tales and stories of Podolia collected and recorded by Mikola Levchenko. Yet other works of great antiquity speak of a total of four moons around Earth. The fourth is our present moon.

Hoerbiger's theory is thought-provoking, but yet another Tibetan chronicle refers to a cosmic mystery which must be considered. This particular Tibetan record refers to a race of giant space beings who arrived on Earth in olden days (remember Chap. 1?); gold-cloth-wrapped bodies of giants are reputed to lie in the crypts of Tibetan monasteries! The giant space beings, of Tibetan legends, were of very high intelligence and ruled Earth and taught Earthlings.

The giant spacemen battled among themselves, then one day they all left in their spaceships when *Mars came close, causing upheaval and geological changes on Earth*! The *Tibetan Book Of The Dead* refers to the spaceships as 'glowing' and 'flying objects'.

Here is another possible indication of a heavenly body instigating geological upheaval on Earth, but *not* upheaval caused by our Moon!

It is not really known why Mu and Atlantis sank. Natural terrestrial changes or geological upheaval induced by nuclear devastation could also explain the enigma. All possibilities should be considered.

Referring again to our Moon. Some proof that the Moon did not separate from Earth, is the 'bulk density'. Earth is 5½ times as dense as water, while the Moon is 3½ times as dense. It is also known that our Moon is by far the largest satellite in proportion to its primary, than any other natural planetary satellite!

Moon rocks brought back by Apollo 11 astronauts were estimated as older than Earth rocks. A US Space Agency report issued 15th September, 1969, said:

> 'It is quite clear that there is a very good chance that the time of crystallization of some of the Apollo 11 rocks may date back to times earlier than the oldest rocks on Earth.'

Also, Lunar rocks contain three mineral arrangements *not found on Earth*! They have been named chromium-titanium spinel, ferro-pseudobrookite, and pyroxmangite.

Perhaps it was *not* chance that captured our Moon, knowing now that it originated elsewhere. Could spacemen, perhaps, have guided our Moon into orbit from deep space – possibly with rays or pulses – for the possible purpose of balancing Earth's orbit around the Sun?

A large quantity of microscopic glassy spherules and similar shapes was found in the 'Moon dust', but nearly 2,500 years ago the Greek philosopher Empedocles (490–430 B.C.) said that the Moon was made of glass! How did he know? Empedocles was the law-giver, physician, poet, and high-priest of Agrigentum in Sicily.

In further reference to giant space beings dwelling on Earth in antiquity, it is interesting to note that Japanese legends which speak of 'fire wheels' descending from the sky, also refer to a race of giant beings with 'supernatural' powers, who mysteriously appeared and took up residence in the Yamaka-Mura and Tatsuyama-Mura mountains at Iwata-Gun.

The men were called Yama-Otoko, and the women, Yama-Uba; the most extraordinary part of this legend is that these beings (space beings?) were reputed to stand about 6 m tall, which is in contrast to Japanese and every other race of Earthmen, who are much less than half of this height!

> 'There were giants in the earth in those days.'
>
> *Gen. 6:4.*

The Chinese, who held great significance for the Canis and Ursa Major Constellations (why?), also have ancient legends of 'giants' dwelling on Earth.

A 1956 excavation of a tomb of the late Eastern 'Chou' – fifth to third centuries B.C., at Ch'u State, Ch'ang-t'ai'kuan, Howan, China – yielded a lacquered wooden figure of a most unusual being. Some characteristics are similar to the Japanese Dogu ceramics. On the face of the Chinese figure are two, very curious, dome-shaped eye coverings surrounded by rings; while on the head, are two short, antenna-like protrusions. Does this figure represent an ancient Chinese god wearing a space suit and helmet with twin antennas?

8. PERHAPS THE 1932 FIND WAS A MUMMIFIED LEPRECHAUN?

'Se non è vero, è molto ben trovato' (*if it is not true, it is a very happy invention*).

(Giordano Bruno, 1585.)

Twelve thousand years ago when Mu was submerged, according to Col. James Churchward, and Atlantis sank, a group of strange, small men arrived in East Asia, apparently from space, and have survived to this day. An oriental archaeologist said:

> 'These people have defied any type of ethnic or racial classification, and the history of their origin is shrouded in mystery.'

To unfold this chapter, some background material is necessary; an editorial in the *Times of India*, 5th January, 1969, said:

> 'There are treatises* which dwell at length on the secrets of laghima or the "power to become weightless attained by developing within each cell of the human body a centrifugal force as strong as the force of gravity contained in the world." '

Also mentioned in this editorial was a 3,000-year-old Sanskrit codex which gave full details for construction of Moon rockets; the manuscript was apparently seized by the Chinese who regarded it as an invaluable source of data to assist them in their own space programme.

It is thought that (possibly) the Chinese may have obtained certain ancient nuclear secrets during the invasion of Tibet; this of course would be distinct from their own research in this field. What possible governing principles of atomic physics known only to the initiated (could) have been preserved in the Potala Palace, Jo Khang Buddhist temple, the Samye Convent, and monasteries Dubung, Sera, and Ganden? Did Tibet preserve

*dated 1000 B.C.

certain abstruse occult mysteries since hoary antiquity? Secrets obtained from where? Lemuria, Mu, Atlantis, Space?

Referring again to the mystery small men and the secrets of weightlessness, it is known that several ancient lands – including Ireland – preserve legends of 'magic' circular levitating 'plates'. It was said that if a certain 'song' was sung to the 'plate' as it was struck, the owner would be able to levitate.

Now Yogis believe that sound vibrations from the human voice have a powerful effect on that person if a certain sound is made – 'mantra' – which corresponds with personal vibrations. Perhaps, in this context, the names of great religious teachers were especially chosen (by spacemen?) as mantras – Buddha, Jesus, Krishna, Rāma, etc. – which would, if spoken reverently by disciples, beneficially influence themselves, the Master, and others?

'Magic plates'; perhaps if a certain mantra is uttered at the same time that vibrations are released from striking the 'plates', the person could levitate?

If the ancients did actually possess such 'magic plates', did they learn construction secrets from spacemen, Atlanteans, etc., or were the 'plates' only for the use of space beings, Atlanteans, etc? Have any of these 'magic plates' been discovered?

It is now well known, from earlier UFO books, that certain unusual discoveries were made in 1938 in East Asian caves; these include 716 mystery, metallic/stone discs, short-statured skeletal remains of beings with huge craniums which cannot be ethnologically classified, and legends relating to these 'Dropa' and 'Ham' – obviously extra-terrestrial – people.

USSR scientists found that the 716 mystery discs, when struck, vibrated at *different frequencies* with a very strange 'rhythm' or 'tone' as if they were electrically charged, or part of an electrical or electronic circuit. The discs also contain an unusually high percentage of cobalt, a white, hard magnetic metal with an atomic weight of 5·90, and other metallic traces.

The *most* notable point about these discs is that they are *exactly* the same diameter – the size of a phonograph record – as the levitating discs described in old legends, and they vibrate at *different* frequencies! This evidently confirms antiquated legends from Ireland, the West Indies, and other places, that the 'magic' levitating plates were cut differently from one another to correspond with each person's individual vibrations!

Are they levitating discs? Were they utilized by the little people of the Dropa and Ham tribes to levitate?

Still other mystery discs have been discovered; the *Cleveland Press*, 8th July, 1968, reported that several hundred strange discs very similar to those described above, were found in 1923 close to Koffiefontein in the Orange Free State of South Africa, 120 km south-east of Kimberley.

William E. Cable, a Johannesburg archaeologist, said:

> 'I found several of these thin stone discs up to seven inches across with a central hole up to two inches in diameter while digging ...'

In the same area were found rock petroglyphs, and graves for tiny beings 90 cm to 1 m 20 cm under the surface of the ground. The estimated age of these discoveries was about 10,000 years, of a similar age to the finds in the central East Asian tableland (i.e. the Mongol-named Bayan-Khara-ula Mountains west of Lake Kuku Nor, between Lat 35° and 38° N. and Long 96° and 100° E.)

Near Bimini, hundreds of cut stone discs about 200 mm in diameter were located washed up in a small sandy quay. Were these once Atlantean levitating discs?

In Fergana (Uzbekistan, Russia), an old rock petroglyph portrays a spaceman clutching a disc about 200 mm in diameter to his chest. On 21st August, 1865 a black, encrusted stone disc fell from the sky at Cashel, Tipperary, in South Ireland. Don't forget that Ireland also has legends of the 'little people' or Leprechauns, Lobaircin, Brownies, Sprites, etc. Were they tiny space beings?

A 1932 discovery may be of considerable significance in relation to the Leprechaun legends. In a small cave at Caspar, Wyoming, an ancient mummy of a tiny man was discovered. The *National Enquirer*, 18th February, 1968, reporting on an X-ray of the mummy by the Anthropological Department of Harvard, said:

> 'An X-ray showed that here was a creature that had been a man, or man-like. Its tiny skull, the spine, the rib cage, the bones of the arms and legs, were readily discernible. The little fellow had been about 14 inches tall in life. Mummified, he weighs about 12 ounces. The X-rays show a full set of teeth. Biologists who have examined it declare that the creature was about 65 years old at the time of death.'

To deliberately mummify a body indicates an advanced race. Does this discovery support the Leprechaun legends?

A September 1956 press report said a Mr. Thomas Hutchinson watched a very tiny UFO land on a Londonderry farm. Was it piloted by a Leprechaun?

The Hindus of old believed there were Sages (Rishis) no larger than a man's thumb. The *Mahābhārata* describes these particular sages – Valakilyas – as: 'resplendent as the Sun, swifter than birds, guards of the Chariot of the Sun'.

9. SPACEMEN WERE IN BABYLONIA 7,000 YEARS AGO

'I came out on the chariot of the first gleam of light, and pursued my voyage through the wilderness of worlds, leaving my track on many a star and planet.'

(Rabindranath Tagore, 1861–1941. From *Gitanjali Song Offerings*, 1913.)

Two ancient documents from the Near East, skilfully translated by scholar of ethnology Y.N. Iban A'haron, are intensely fascinating. The first, a 7,000-year-old work, the Hakaltha or 'laws of the Babylonians', states:

> 'The privilege of operating a flying machine is great. The knowledge of flight is amongst the most *ancient* of our heritages, a gift From Those Upon High. We have received it from Them as a means of saving many lives . . .'

The second document, a 5,000-year-old Chaldean work – *Sifr'ala* – actually gave instructions to construct a 'flying machine'. Y.N. Iban A'haron translated the following words from this document: 'calibration', 'crystal indicator', 'copper coils', 'equilibrium', 'gliding capacity', 'stability', 'wind resistance'.

According to this document, the 'crystal indicator' changes colour during flight.

Many years ago, the archaeologist Dr. J. O. Kinnaman, together with the famous Sir William Matthew Flinders Petrie (1853–1942), excavated in Egypt what seemed to be component parts of the legendary Hebrew 'flying machines'. Also discovered was a 3,500-year-old metal insignia, apparently worn by pilots of the 'flying machines'.

The Albright-Knox Art Gallery in Buffalo, New York State, possesses an exquisite copper figurine of an ancient god, of features although human, so noble and unusual, that it resembles no known race on Earth. Spectroscopic analysis indicates that the figurine is at least 5,000 years old; it was discovered in an antique store in Baghdad in 1951, but its origin is a mystery. The most logical origin, according to art experts, is the Tigris-Euphrates Valley, but the experts consider the figurine to be such a masterpiece, that they know of no ancient civilization sufficiently advanced to be able to produce it. The features of the figurine resemble those of no known Earthman, ancient or modern!

Who was this ancient god? Was the figurine made by spacemen and left on Earth, or was it made by Earthmen with a civilization much more advanced than we are aware of? If so, does the figurine resemble one of their gods who taught these ancient Earthlings their advanced sciences?

Recent discoveries in the Tigris–Euphrates Valley include *circular discs* of baked clay portraying *serpents* and *stars*!

An article in *Sunrise* for October 1969, referred to the ancients of the Tigris–Euphrates Valley, when it was remarked that:

> 'We cannot help but marvel that these ancient peoples could predict eclipses of the sun and moon without our specialized instruments, chart the synchronous movements of the heavenly bodies, and clock the passage of time with such infallible accuracy.'

It is known that the ancient Assyrians possessed optical lenses, but even this does not explain their advanced astronomical knowledge. From whom did they learn?

The Babylonians had a pantheon of 'seven planetary gods'. The Hebrew *Book of Enoch* speaks of 'angels' from the 'seven heavens'; each of these 'angels' was in command of 496,000 other 'angels' – or space beings?

A Nineveh discovery was ancient clay tablets depicting Gemini and Scorpio; the ancient Assyrians were proficient not only at astronomy, but also astrology. Although astrology is disbelieved by some, these following facts validate this ancient science:

(A) Some Indian universities require a twelve-year study course before an astrology proficiency degree is granted.

(B) A definite link between electromagnetic storms and movements of the heavenly bodies was ascertained by John Nelson, a top US radio meteorologist.

(C) Still another definite link between culminations of the planet Uranus, and earthquakes, was established by German geophysicist Rudolph Tomaschek.

(D) At the US Northwestern University, Dr. Frank Brown discovered a correlation between planet cycles and plant, as well as animal, metabolic rates.

(E) Neuro-psychiatrist Dr. Leonard Ravitz observed a relationship between the electro-magnetic potential of the human body and solar/lunar cycles.

IN THE FIRST CENTURY A.D. JOSEPHUS DESCRIBED THE UFO ADVENTURES OF MOSES, CIRCA 1500 B.C.

Let us first define the meaning of the name Moses, and related to his name, the origin of the alphabet. Moses is an Egyptian name meaning 'Thutmose' or 'Thoth's child'. In Egypt of old, Thoth was known as the god who taught writing, mathematics, geometry, and astronomy. Probable confirmation for this legend is that the alphabet from which over 200 other alphabets are derived, came from the desert region somewhere between Egypt and Babylonia about 3,500 to 4,000 years ago. This apparently explains the story of Thoth, but Sumerian cuneiform dates back about 5,500 years, while the Chinese say the alphabet was 'invented' by the god T'sang Chien who travelled in a 'flying dragon' with 'four eyes' – a spaceship with four circular windows?

Although recorded history dates back only a few thousand years, very archaic legends say that the ancients once recorded all knowledge on 'tablets' of rare metal or crystal; memory units for computers? This system used extensively on Mu and Atlantis, according to legend, was eventually lost to Man and replaced by writing.

Referring again to Moses; during the reign of Rameses II (Osymandias), Moses witnessed many strange sights. One of the most unforgettable incidents in the life of Moses was the probable meeting – or meetings – between himself and cosmic entities. Did Moses see a UFO or UFOs on Mt. Sinai? (Probably the North Peak of the Jebel Musa Range on the Sinai Peninsula.)

A fascinating and accurate account of the first and a later visit of Moses to Mt. Sinai, is in the Works of Josephus; the account of the first visit of Moses to Mt. Sinai (from *Antiquities of the Jews*, Books I–VI, Chapter XII), states in part:

> 'And here it was that a wonderful prodigy appeared to Moses: for a fire seized upon a thornbush, yet did its green luxuriance continue intact, nor did the flame consume the bush, though the flame was great and fierce. Moses was frightened at this strange sight, but was still more astonished when the fire uttered a voice and called to him by name, and spoke to him, and told him how bold he had been in venturing to a place where no man had ever come before, because the place was divine; and advised him to remove as far as possible from the flame, and to be contented with what he had seen . . .'

Did the bush not catch fire due to a glowing UFO, and *not* to an actual flame? Why did the 'angel' advise Moses 'to remove as far as possible from the flame, and to be contented with what he had seen . . .'?

Was there danger of physical injury from the 'force-field' of a UFO? There are twentieth-century reports of radiation danger, and of people burned by approaching too close to landed UFOs.

When 'the fire uttered a voice', was it, in reality, a spaceman speaking to Moses electronically through a loudspeaker on the UFO?

A later visit of Moses to Mt. Sinai is described in Chapter V of *Antiquities of the Jews*; the Hebrews were awaiting the return of Moses with an 'oracle':

'So they passed the days in this feasting, and on the third day, before the sun rose, a cloud* spread itself over the whole camp of the Hebrews, such as none had ever before seen, and encompassed the place where they had pitched their tents: and while the rest of the sky was clear, there came strong winds that raised up large showers of rain, which became a mighty tempest. There was also such lightning as was terrible to those who saw, and thunderbolts hurled down declared God's gracious presence and favours to Moses. Now as to these circumstances every one of my readers may think as he pleases, but I am obliged to relate this history as it is described in the sacred books. What they saw and heard frightened the Hebrews terribly, for it was such as they were not accustomed to; and then the rumour that was prevalent, that God frequented that mountain, greatly awed their minds, so they sorrowfully confined themselves within their tents, supposing that Moses was destroyed by the divine wrath, and expecting the same for themselves.

'As they were under these apprehensions, Moses appeared majestic and in great elation. His appearance freed them from their fear, and suggested better hopes of what was to come. . . .'

Was Moses brought down to the Hebrew camp by UFO?

UFO PROOF IN THE DEAD SEA SCROLLS AND APOCRYPHAL WORKS?

It is known that ancient manuscripts have, on many occasions, been destroyed by peoples of other faiths if the data contained by them do not conform to their, or orthodox, beliefs. Some historical examples are already known to some people: one instance was the burning of

*'cloud', or UFO?

80,000 manuscripts from classical Arabic authors at Granada by Cardinal Ximenes. It is said that 'whole passages' of rare ancient treatises in the Vatican libraries have been erased or blotted out.

What of the Dead Sea Scrolls, and apocryphal works? The *Divine Chariot* for example describes the descension and ascension of a heavenly vehicle. Some descriptive words in this record, are: 'when the wheels turn' and 'ascend'. The flying 'wheels' are described as the colour of 'bronze', with an aura or emanation of 'fire'. They are further described as a combination of several colours; living beings within the 'wheels' are said to have worn beauteous 'shining' or 'brilliant' raiment, appearing to emanate colours of exquisite radiance. Like a shimmering metallic space suit?

An interesting Biblical observation, are the Seraphim (*Isaiah, 6:2*). Were they spacemen with six-bladed, back-pack helicopters?

WAS THE STAR OF BETHLEHEM A SPACE-SHIP?

Belief in a 'lucky star' originated with the Star of Bethlehem. The birth of Christ has been fixed by three Italian professors at 6 B.C.

A most unusual painting, said to be one of seven painted by St. Luke, can be seen near Madras in the Church of Our Lady of Expectation on St. Thomas' Mount.

The painting portrays the Madonna and Child with a puzzling egg-shaped Star of Bethlehem in the upper left corner. Within the elliptical-shaped outer covering is a small, round, light-coloured central section which is encircled by a larger dark round area emitting rays. A rod fastened to the circular dark section protrudes a short distance out of the back of the egg, and this is framed on each side by wavy-shaped appendages (wings?). Does this painting depict a spaceship?

Herod (73–54 B.C.), the king of Judaea, said in reference to the Star of Bethlehem:

'The Great Star had risen before his conversation with the Maji. Now it flamed overhead *again*.'

(my italics)

An apocryphal work states:

'So the wise men went forth, and behold, the Star which they saw in the East went before them, till it came and *stood* over the cave. . . .'

(my italics)

Herod said the 'Star' twice 'flamed overhead'; the apocryphal description said the 'Star' 'stood over the cave' (see also *Matt. 2:9*). No star could 'flame' overhead twice or stand still. Was it a spaceship? Was the Christ-child brought to Earth from another planet?

My discovery of the apocryphal description of the Star of Bethlehem was from the Protevangelion by James the Lesser, the first Christian Bishop of Jerusalem. In the Siberian journal *Baikal* and the Soviet digest *Sputnik No. 1*, Vyacheslav Zaitsev – Master of Philology, Soviet Byelorussian Academy of Sciences – describes another amazing apocryphal work also referring to the strange 'star'.

Zaitsev says that he has 'irrefutable historical proof' that the Star of Bethlehem was a spaceship. He studied a fifteenth-century Byelorussian translation from Latin apocrypha, which states that the Star of Bethlehem 'had wings, the kind an eagle has, and many long rays'. The 'star' hovered over Mt. Vans in Turkey (then part of Armenia, Lat 38¾° N., Long 43° E.) for an entire day, then flying in circles, 'alighted on the mountain like an eagle'.

'Certain books' in the Apocrypha say that Christ descended from the 'star', which was described by Palestinian nomads as a 'flying temple', or possibly dome-shaped!

The arrival of the infant Christ on Earth from a spaceship is less fantastic, more credible, logical and acceptable, than the ethereal dogma taught by the Christian Church. This *does not* imply, or denigrate, Christ as a being lesser than divine, and supernatural spiritual powers attributed to Him in historical works – particularly the newly-discovered Arabic *Testimonium Flavianum* extracted from Josephus' writings – must undoubtedly be true.

Of interest in the extra-terrestrial origin probability for Christ is the Holy Shroud of Turin which shows the impressions of a man, and is thought by many erudite scholars to be the actual winding sheet of Christ. Italian professor and sculptor, Lorenzo Ferri, basing his conclusions on the 'Shroud' impressions, measurements, and a plaster figure made by him from the 'Shroud' impressions, etc., proved that the 'Shroud' figure, if Christ, was 1 m 85 cm tall. During Christ's time the average height of Palestinian men was only 1 m 60 cm tall! From *whence* did Christ come?

Most of the following persons were said to be 'divine', and *all* were associated with a huge 'star' which performed manoeuvres most uncommon for a celestial body:

Abraham	
Aesculapius	(a 'serpent' being, the Greek god of medicine).
Bacchus	(or Dionysus, the Roman and Greek wine god who taught fruit cultivation and wine production).
Buddha	
Krishna	(the Hindu deity).
Confucius	(551–478 B.C., the Chinese sage).
Jesus	
Mithras	(the Persian sun-god).

Romulus	(the co-founder of Rome).
Socrates	(470–339 B.C., the Greek philosopher).
Zoroaster	(the Persian mystic).

Legends of antiquity say that before Zoroaster was born in 660 B.C. a strange 'light' *remained stationary* over his village for the entire period of 'three days and nights'. It may be more than coincidental that in 630 B.C., when he founded Zoroastrianism, he associated the 'winged disc' symbol with his religion.

The Apocrypha mentions 'lights from heaven' containing angels. Buddha and Krishna were said to have ascended to 'Heaven' in a flying 'light' when their teachings on Earth were completed. Mohammed and Zoroaster also ascended to 'Heaven'. Were some of the great religious teachers of old associated with space entities? Flying 'lights' and stationary 'stars' – were they UFOs?

THE TRANSFIGURATION

Did Jesus enter a UFO on Mt. Tábor in Galilee, North Palestine, 11 km 263 m south-east of Nazareth, and converse with two spacemen, the reincarnated souls of Moses and Elijah?

Jesus led Peter, James and John, up Mt. Tábor, where several highly unusual incidents occurred. The first was the sudden transfiguration (changing the outward appearance) of the raiment of Jesus.

Mark 9:3 states:

> 'And his raiment became shining, exceeding white as snow; so as no fuller on earth can white them.'

Matt. 17:2 states:

> 'And he was transfigured before them: and his face did shine as the sun, and his raiment was white as the light.'

Did Jesus change His garments for a shimmering, metallic space suit and helmet?

After the transfiguration of Jesus, Moses and Elijah appeared and conversed with Jesus. While conversing, a 'bright cloud' (*Matt. 17:5*) overshadowed the group and terrified Peter, James, and John; was the 'bright cloud' a UFO? The disciples fell on their faces with fear, as the 'bright cloud' also spoke to them (through a loud-speaker?). After reassurance from Jesus that there was nothing to fear, the disciples rose, and saw that the 'bright cloud', along with Moses and Elijah, had disappeared.

While coming down from the mountain, Jesus requested the disciples: 'that they should tell no man what things they had seen. . . .' (*Mark 9:9*).

While discussing the transfiguration and the 'bright cloud' among themselves, the disciples wondered why Elijah, or Elias (and Moses?), were conversing with Jesus, as Elijah had lived on earth several hundred years earlier. The disciples asked Jesus:

> ' "Why then say the scribes that Elias must first come?" And Jesus answered and said unto them, "Elias truly shall first come . . . But I say unto you, That Elias is come already, and they knew him not . . ." Then the disciples understood that he spake unto them of *John the Baptist*.'
>
> *Matt. 17:10, 11, 12, 13.*
> (my italics)

Did Jesus mean that John the Baptist (a spaceman?) was the reincarnation of Elijah, who, incidentally, went up 'into heaven by a whirlwind' (*II Kings 2:1*) (a UFO?) over 900 years earlier! Christ did not directly teach reincarnation, but apparently did refer to the subject. The Essenes, and Josephus, spoke of reincarnation as an accepted fact.

A painting from the 1538 Venetian Minehya portrays the 'bright cloud', which bears no resemblance to a cloud, on Mt. Tábor, with Jesus standing within the 'cloud'.

The 'cloud', which is pictured as oval, is dark in the centre, with a lighter-coloured surrounding edge, which seems to be emitting rays around its circumference. Did Jesus enter a UFO on Mt. Tábor and possibly take a flight, converse with, and receive instructions from (?) two spacemen? An arrow head superimposed over the oval definitely indicates that the 'cloud' is about to fly vertically up into the sky!

THE BURNING LIGHT ON THE MOUNT OF OLIVES

One of thirteen Coptic manuscripts inscribed nearly 2,000 years ago by Egyptian Gnostics, and discovered by Arabs north of Aswan in an ancient cemetery in 1946, preserves most singular knowledge. This ancient manuscript, now kept in the Cairo Coptic Museum, recounts that Christ resurrected, appeared to the apostles on the Mount of Olives as a 'burning light' – or was it a UFO?

THE HEAVENLY THRONE

In *Revelations 4:3, 4, 5, 6* a heavenly 'throne' of unusual aspect is described:

> 'And there was a rainbow about the throne, in sight like unto an emerald. And round about the throne were four and twenty seats: and upon the seats I saw four and twenty elders sitting, clothed in white raiment; and they had on their heads crowns of gold. And out of the throne proceeded lightnings and thunderings and voices. And before the throne there was a sea of glass like unto crystal: and in the midst of the throne, and round about the throne, were four beasts full of eyes before and behind.'

The twenty-four 'elders' with 'white raiment' and 'crowns of gold', were probably beings wearing space suits and helmets. The 'sea of glass' or 'crystal' which was before the 'throne', was apparently the transparent, or translucent, material of a UFO, while the 'four beasts full of eyes' were circular windows. The 'lightnings and thunderings' were undoubtedly the light, and sounds, made by the UFO propulsion system, while the UFO also had a green corona, i.e. 'emerald' 'rainbow' surrounding the 'throne'!

THE CHOSEN PEOPLE

Why is the Jewish race referred to as 'the chosen people'? Why does their history date back only a few thousand years, while other civilizations are seemingly many thousands of years older (Bahamas, etc.)

Were the ancestors of the Jewish people a *select group* of space-being settlers brought to Earth only a few thousand years ago? A British scientist believes that, genetically, the Jewish race is superior to every other race on Earth.

In 1969, archaeologists discovered in the Qaf'zeh cave, close to the Mount of Precipitation in Galilee, an archaic grave containing the 50,000-year-old skeleton of a ceremonially-buried child – *apparently Homo sapiens* – which indicates that civilized people were living in the area *many* thousands of years before the Jews arrived.

WHAT DOES THE 'SCHEMA' SAY?

The Roman Catholic Church is interested in the probability of universal life, and has prepared a 30,000-word 'schema' giving serious consideration to the distinct possibility that Earthmen may find intelligent beings on certain planets of our solar system during future space flights.

Monsignor Roberto Masi, Rector of the Roman College for Juridical Studies, said:

> 'St. Thomas Aquinas noted that, in *Genesis*, the Creator made the sun, moon and stars for man but that did not exclude that they could benefit other beings. Nor has Catholic thinking, since the accept-

ance of Copernican theories of the universe, ever rejected the thought of life on other worlds.'

A TERRESTRIAL WORLD GOVERNMENT OF SPACEMEN?

Why do certain religious groups believe in a second coming of Christ, and, or so it seems, a World Government ruled by angels? A time which is supposedly rapidly approaching.

Are Biblical records the result of computer estimates programmed by spacemen, which indicated that it would take approximately 2,000 years for Earth's population to reach such a density, that total self-annihilation due to nuclear and biological/chemical warfare, would be a very real possibility?

Did Nostradamus and Edgar Cayce prophesy possible nuclear/biological/chemical war between A.D. 1998–2000 which would be averted or stopped by spacemen, who might rule the world for our *own* protection?

10. PHAËTHON CRASHED A UFO INTO THE RIVER PO

'Phaëthon rode the sun, and here is his tomb.
His daring was the reason for his doom.'

('The Phaëthon and Photons', by Andrew Thomas, *Australian Flying Saucer Review*, June 1965.)

Greeks of old referred to other planets as: 'the houses of the gods'. There was a time, according to legend, when the gods walked with men, solved their problems, and took part in their battles. Some people were considered very fortunate if they married space deities; their children were thought of as 'half-divine'.

This same race of space beings was known in India of antiquity – i.e. the Vidyā Dhara (male) and Vidyā Dhari (female) space beings, who married Earthlings to produce a superior genetic strain among their offspring.

Ancient Greeks, like the Hindus of antiquity, did not regard the gods as myths, but as real beings who descended from the skies; they were said to exactly resemble Earthlings.

If our ancestors were space beings who migrated to Earth in ages past, can we really be certain that new-arrival extra-terrestrials are not walking our city streets today – unnoticed?

The Cyclopes of Greek legends were believed to be the offspring of a union between an Earthborn parent and a space being. The Cyclopes who dwelt on Mt. Etna in Sicily, were said by Homer to have been giant shepherds who associated with the gods. Their king was Polyphemus. The Cyclopes were also known as Titans, i.e. children of Gaea (Earth) and Uranus (Heaven).

In Corinth, a temple was constructed to worship the Cyclopes. It might only be mythology, but in a different part of the world, in ancient Peru, a pre-Inca race has left rock petroglyphs, depicting on one a celestial being of giant stature *with one eye in the centre of the forehead*!

What does this mean? Was there an ancient link between a pre-Inca race and the ancient Greeks? Were the Cyclopes spacemen, half-spacemen, or what? An obsolete one-eyed bodily form – or living robots?

DID THE GREEK GODS RULE IN EGYPT?

The first book of Manetho, the Egyptian priest and historian of Sebennytus, states that the god Kronos (Saturn) ruled as a king in Egypt for 40 years and 6 months; Apollon (Apollo), a demi-god (half-divine), ruled 25 years in Egypt; Ares (Mars), another demi-god, ruled in Egypt 23 years; Zeus (Jupiter), also a demi-god, ruled in Egypt for 20 years.

Zeus was a demi-god but apparently inherited certain privileges from the gods, which included periodic transport to the heavens (in spaceships?). Zeus and other deities were said to live on Mt. Olympus between Macedonia and Thessaly; from Mt. Olympus they were believed to ascend to the heavens! Ganymede was considered as 'the most beautiful of mortals' and was taken away by Zeus on an 'eagle' (UFO?) to live among the gods.

At Ephesus, an ancient city in western Asia Minor, founded approximately 1,100 years before Christ, the ancient Greeks worshipped a statue of Artemis (Diana) believed to have been brought from Heaven by Zeus. The ruins of Ephesus still remain and are situated on the Cayster river bank, south-east of Smyrna or Izmir in Asiatic Turkey.

The statue is even mentioned in the Bible:

> 'Ye men of Ephesus, what man is there that knoweth not how that the city of the Ephesians is a worshipper of the great goddess Diana, and of the image which fell down from Jupiter? Seeing then that these things cannot be spoken against, ye ought to be quiet, and to do nothing rashly.'
>
> (*The Acts 19:35, 36*)

Did Zeus bring the statue of Diana to Earth from another planet? The statue is now kept in the Capitoline Museum, Rome.

On the West Terrace at Nemrud-Dagh in Turkey's Anti-Taurus mountains, a forgotten civilization which existed about 2,000 years ago has been uncovered by archaeologists. Among the ruins are many huge heads – as high as a five-storey building – of Greek gods carved in stone with interesting headdresses; some are conical, while others are rounded on top. Stylized representations of space helmets?

PHAËTHON CRASHED THE SUN-CHARIOT

The most significant Greek legend from the point of view of a possible UFO association is the story of Phaëthon (Greek: 'The Shining').

It was once said that Phaëthon asked his father Phoebus for a chance to drive the 'Sun-Chariot' which was powered by 'fiery horses'. Phoebus attempted to dissuade his son by saying that the Sun-Chariot was *far* too big and powerful and quite dangerous. Only the gods could control tremendous power.

On the insistence of Phaëthon, Phoebus reluctantly took his son to the coveted Sun-Chariot, which in the light of day was gleaming, as if made from the finest gold, silver and polished gems. Phoebus spent some time instructing his son how to fly the Sun-Chariot and cautioned him to control the 'fiery horses' most carefully as they were difficult to hold. He warned Phaëthon to navigate with great care through the five heavenly zones, i.e. Taurus, Sagittarius, Leo, Scorpio, and Cancer.

Phoebus then placed a special 'crown' (space helmet?) on the head of Phaëthon and sadly bade him farewell. The worst fears of Phoebus were realized when out in space Phaëthon completely lost control of the Sun-Chariot, due to his great fear and inability to hold the 'fiery horses'. He hurtled back to Earth, causing extensive damage to Terra due to the 'fiery breath' of the 'horses'. Zeus was forced to fire a 'thunderbolt' (missile?) at

Phaëthon's aerial 'chariot' and bring him down before even more damage was caused from the 'fiery horses'.

Phaëthon and the 'chariot' came to a sad end; both crashed into Eridanus (the River Po) in Italy which irrigates the plains of Lombardy and Piedmont.

Was Phaëthon an inexperienced pilot attempting to fly a UFO? Was the damage effected on Earth the result of lack of control over atomic, or photon, motors ('fiery horses')? And Zeus' 'thunderbolt', was it a missile (atomic?) which he was coerced to use to save Earth from catastrophe?

DID ALEXANDER THE GREAT TRAVEL IN A UFO?

The ancient Greeks closed their dramas with the appearance of a 'Deus ex Machina' – 'a god from a machine' who solved all problems.

Of interest in this connection is the old story of the flight of Alexander the Great to the 'heights of heaven' on the back of an 'eagle' (UFO?) where he was able to study the topographical features of Earth and plan his world conquest. Other, similar mysteries from ancient Greece are, for example, as follows.

It was recorded that when Apollo was born 'swans' came from Pactolus and circled around Delos seven times. Were the 'swans' UFOs? Strange 'silver swans' are described in the Hindu epic *Rāmāyana*!

Prometheus, the son of the Titan Iapetus, and Clymene, is described in Greek legends as having brought the gift of fire from 'Heaven' in a 'hollow tube'. An electro-mechanical, gas, etc., method of artificially producing fire? Prometheus was also credited with being a great teacher of men. Was he also the Maori deity Auahi-turoa who brought the gift of fire?

Did space beings once dwell in ancient Greece, helping and teaching the inhabitants agriculture and science, marrying them and raising their 'half-divine' children?

Among the inventions of the Greeks and Romans were an alarm clock and a working steam engine. Some of their inventions were far in advance of their time. (?)

A mystery from a later period is that at Novgorod on the Upper Volga a fourteenth-century lead finger ring once belonging to a Greek intellectual was discovered with 'coded inscriptions' referring to a certain Russian deity. Byzantine-Greek Christianity at that time was well established on the Upper Volga. What significance did the Russian god have in the life of this Greek?

11. ARE SPACEMEN CONCERNED WITH DEGENERATED NEANDERTHALS, STILL SURVIVING IN THE CAUCASUS AND NORTH AMERICA?

'The number of inhabitable systems is about 3–5 per cent of the number of stars; this leads to eight billion inhabitable systems in our galaxy.'

(Dr. Su-Shu Huang, Dearborn Observatory, Michigan)

Possibly one of the oldest mysteries from the Americas are US and Canadian Indian legends of gigantic man-like beings covered in black or red hair, and known as Bigfoot and Sasquatch.

In recent years these creatures have been sighted as frequently as in past centuries, and seem to have lived in the Americas since far back in pre-history.

Several plaster casts of Bigfoot footprints have been taken and are found to resemble, to a degree, those of Man, but range in size from those of youngsters to approximately 42–56 cm in length.

On 20th October, 1967, a Mr. Roger Patterson managed to obtain a few feet of 16-mm colour movie film of a Bigfoot alongside Bluff Creek, in the largely unexplored, desolate wilderness of the National Forest, Northern California.

Scientists from the Smithsonian Institution in Washington, the US Coast and Geodetic Survey and Emery University, the University of Alberta, Edmonton, Canada, and University of Calgary, Alberta, have all examined enlarged frames from the film. Having seen this film sequence myself, I cannot agree with scepticism, and also these beings are sighted in many areas of the world: USA/Alaska, Canada, Africa, Himālayas,

Pamirs, Caucasus, Central Asian deserts, Mongolia, Tibet, Chinese Turkestan, and Malaysia!

A striking creature, similar, or identical, to the American relative Bigfoot, is seen in Russia where it is called Meshe-Adam in Azerbaijan, Agach-Kishi in Karachai, Tkiskasti or Shaitan in Georgia (Caucasia), and Mazyl or Almasti in Kabardinia (North Caucasus). All these names mean 'forest' or 'wild' man. As in the Americas, the Russian Bigfoot cousins are covered in black or red hair.

The creatures are sighted in the Tien-Shan mountains on the Northern Frontier of Chinese Turkestan, while the Mongolians call them Almas. Of course there is the enigmatic Ye-Ti or Metch-Kangmi of the Himālayas, while the Maoris preserve their legends of the Maeroero, the ferocious being or beings covered in long hair who once lived in the mountains. If the Maeroero was a Bigfoot or related to it, how did they appear in New Zealand, so far from the rest of the world?

Supporting my Chapter 1 theories was an interesting statement from Dr. Boris Porshnev of the USSR Academy of Sciences, who referred to the 'wild hairy man' when he remarked:

> '. . . in this, it was realized that it was a problem not of an ape, but of a relic hominoid, of Neanderthal man who had degenerated, and has survived through all the millennia of human history.'

Dr. Porshnev also remarked that the dried wrist of a Himālayan Ye-Ti, and Ye-Ti and 'wild man' footprints, are anatomically identical in appearance with Neanderthal Man's footprints and wrist.

Of course Bigfoot is larger than Neanderthal Man, but this can be explained within my hypothesis of *further degenerative genetic mutation* of artificially created biological robots, and a seeming gradual reversion to apes!

US evolutionists and others say that no bodily remains of a Bigfoot have been found. This may or may not be correct with regard to the Americas, but in the Soviet Union documented cases do exist of such finds.

The following detailed word-for-word report was written down on 30th June, 1964 at Kabardinia, by J. I. Kofman and O. L. Opryshko, special correspondents of *Kabardinio-Balkarskaya Pravda*. It may also be noted that biologist Janna Kofman is a Member of the Geographical Society of the Academy of Sciences of the USSR.

The verbal description recorded here was from an interview with K. K. Zhigunov, 46, a worker at the Kabardinian Baksan Brickworks.

> '. . . I took a short cut across a maize field. About 40 paces away from the road I found the remains of an "almasti". Wolves, or dogs, had been at work on them. On an area of about 12 metres in diameter the maize was broken and trampled. In the middle lay the head of the almasti with part of the neck. The left half of the neck was eaten away. Until then I had not believed in the existence of almastis and, therefore, inspected the head. With a stick I turned it over and, squatting, I took a careful look. The head was covered with a crop of very thick and long hair, which was matted and plastered with burdock. The hair was so thickly matted that when I turned the head over the hair made a pillow for it. Because of that I could not determine the shape of the skull, but I would say that it was about the size of a human skull. It had a somewhat sloping forehead, a small snub nose without a bridge like that of an ape, and prominent cheek bones. The lips were thin and straight like an ape's. I did not see the teeth. The lips were caked and I did not open them. The chin was round and heavy, not like a man's. The ears were like a man's. One ear was torn. The eyes slanted downwards. I don't know the colour – the eyelids were closed and I did not open them. The skin was black and covered with dark chestnut hair. There was no hair round the eyes and on the upper part of the cheeks. There was short hair on the cheeks and ears, and longer hair on the neck.

'A pungent odour rose from the head. It was not

the odour of decay, because the remains were fresh and did not smell. There were no flies or worms – evidently it had been brought down only some hours before, and the blood was newly caked. It was the odour of the almasti itself, and it was so nauseating that I nearly vomited. I therefore held my nose with my left hand and had a stick in my right when I examined the head. It was the odour of old mud, of an unwashed body, of mould.

'Other parts of the body were scattered nearby. I saw bones covered with remains of dark flesh, but I did not go near them, and did not look at them.'

It is of considerable interest that UFOs have frequently been sighted in the Caucasus and California.

Possible or probable confirmation of interest by spacemen in Bigfoot, might be found in the legends of the Ingushes, Kabardinians, Chechens, Georgians, and Karachais of Russia; they believe that these creatures are 'evil spirits' which once tried to 'climb to heaven' and are being 'punished by Allah'.

Are they 'fallen angels' – spacemen perhaps from a different planetary environment who have degenerated by terrestrial atmosphere and vibrations, or by deliberate drug- or radiation-induced genetic mutation, or ... are they degenerated biological robots – Neanderthals – created by spacemen or Earthmen in antiquity? (Seemingly the most likely possibility?)

A new law announced in December 1970 protects Bigfoot from hunters in Washington State by imposing a five-year jail term and an $8,000 fine on any offender. It is obvious that authorities believe in the existence of the Bigfoot/Sasquatch!

QUETZALCÓATL WAS A SPACEMAN

Nearly 2,000 years ago a god travelled the Americas, teaching about the Supreme Deity, healing the sick, and persuading the people to desist from sacrifices. According to a Toltec legend this god was a being of fair countenance and aspect, a gentle soul of benign manner.

Tahitian legends say that he first appeared in the islands from a swift fleet of 'great birds' with 'great wings' (UFOs?) while the Toltecs claimed that he came from a 'hole in Heaven' in a 'winged ship' and landed at what is now known as Vera Cruz in Mexico.

This was the divine QuetzalCóatl, the founder of Mexico, especially worshipped at Cholula, although he journeyed the Americas, teaching healing, and bestowing the gift of seeds – a 'gift from the gods'?

A Red Indian legend says that he once arrived in a 'fire' (UFO?) from the sky, and was clothed in 'glowing' garments (a spacesuit?).

A Toltec legend of old relates that QuetzalCóatl not only taught of the Supreme Intelligence, but he also taught astronomy, medicine, and agriculture. The cedar was sacred or 'holy' to QuetzalCóatl. Was it brought from another planet?

After a lengthy period teaching and healing, QuetzalCóatl departed from Cozumel Island in the Yucatan, south-east Mexico, in a shining 'serpent' vessel (a spaceship?) to the land of Tla-Pallan; he promised to return but was never seen again.

Other legends relate that QuetzalCóatl was always interested in visiting Venus, and finally departed to that planet!

QuetzalCóatl was, it seems, one of the serpent people from probably a planet of Barnard's Star or 70-Ophiuchi!

The name QuetzalCóatl is derived from the word *quetzalli* or 'green feather'; this is further linked with Quetzal – *pharomacrus mocinno* – the Mexican bird. Cóatl is derived from *cohuatl* or 'snake'. So his name really means 'feathered serpent'.

THE INSCRUTABLE BEINGS IN WYOMING

On the summit of the 3-km-high Medicine Mountain in Wyoming is The Medicine Wheel, an ancient circle of stones 22 m 20 cm in diameter, with 28 spokes radiating out from the centre. Even more odd, is that one of these spokes – also made from stones – radiates out 2 m 70 cm beyond the rim. At uniform distance around the rim are spaced five cairns; four of these stand 75 cm high, except one – the largest, which faces the rising sun. The centre cairn measures 3 m 60 cm in diameter, by 90 cm high.

The Medicine Wheel was discovered in 1887 on this mountain plateau, so difficult to reach by modern means that it must have been almost completely inaccessible for ancient Man, yet someone was here in olden times!

The Medicine Wheel is an Indian name, but the builders of the Wheel cannot be positively identified. Its purpose has been the subject of much controversy but indications seem that it was once used as a primitive but efficient astronomical computer.

The most curious observation about the plateau is that several holes have been discovered; when a rock is dropped into one of the holes it can be heard bouncing into the mountain for 60 seconds. Writer Robert B. McCoy speculated whether the holes were once 'ventilator shafts' and whether the mountain might be 'hollow'!

Did advanced alien beings (from where?) hollow out the mountain for a subterranean dwelling? (Remember the tiny mummy found in Wyoming?)

BUDDHA AND/OR HIS DISCIPLES IN NORTH AMERICA?

Another curiosity in North America is the intriguing possibility that Buddha and/or his disciples once visited Alaska, Montana and Idaho!

The 'Feet of Buddha', a symbol extremely sacred in the Orient is a footprint with 65 symbols associated with Buddhism. Some of these symbols are: a conch shell, two fishes possibly associated with the Piscean Age, the 'Wheel of the Law' or 'Wheel of Doctrine', and a Swastika on each toe. The Swastika symbol has been discovered in many parts of the ancient world. Apart from the Buddhists, the following peoples of antiquity used this symbol: Brahmins, Chinese, Cypriots, Greeks, Japanese, Persians, Scandinavians, and Tibetans.

Swastika, a Sanskrit word, was a symbol associated with several – particularly celestial – interpretations. Still another theory is that the Swastika, known as far back as the fourth millennium B.C., symbolizes the four convergent rivers of 'paradise'.

An intriguing discovery are archaic petroglyphs in North America portraying Buddha's footprint with some of the oriental symbols. The most fascinating depictions were found on Hope Peninsula on the shore of Lake Pend Oreille, Idaho.

Was Buddha and/or his disciples in North America thousands of years ago? If so, how did they arrive?

THE WHALE-CARRYING FLYING MACHINE

Red Indians of the north-west coast knew of a gigantic flying machine which passed most of its time under water, but occasionally it would rise from the water and fly away while flashing 'lightning' and making 'thunder' as it went. It was so huge, according to legend, that it was capable of carrying a whale!

Red Indian legends of the underwater spaceship or spaceships are closely paralleled by the Roc – Rokh – Rukh – mentioned in the *Arabian Nights*, and the Hakuwai in Maori legends!

According to legend Sindbad was believed to have been taken off by a Roc, a bird so gigantic that it was supposedly able to carry three elephants!

Marco Polo said the Roc lived in Madagascar. Fossil bones and eggs of a huge bird – Aepyornis – have been found in Madagascar, but it is unlikely to be the Roc in the *Arabian Nights*!

Maori legends refer to the Hakuwai, a gigantic mythical bird of the heavens. It is not possible to fix the origin of the Hakuwai from the legends. It came from either Antares, Sirius, the twelfth heaven, Jupiter or Mars.

Was the underwater spaceship known to the Red

Indians, the Roc and the Hakuwai, the same type of underwater spacecraft?

What was probably a massive underwater spaceship was seen in recent times and studied by USSR scientists! In 1964 Soviet scientists were investigating their own underwater puzzle in Lake Vorota, in the Sordongnokh Plateau, Siberia. This inexplicable 'sphere' was a huge unknown shape, which moved about on the surface and underwater.

Moscow News reported that a Soviet geologist who had seen the solid shape described it as resembling:

> 'an ominous looking dark grey sphere. Its body resembled an oversized glistening tin barrel with a slanted horn on its back.'

Moscow News further reported:

> 'As the monster came closer, the geologists saw two bulging protrusions resembling eyes, spaced about two metres apart.'

It was also announced that expeditions of scientists and geologists from Moscow had seen the 'strange spherical shape' but were unable to identify it. Suggestions that it was a prehistoric beast, or a whale, were discounted due to the extreme improbability of any creature surviving in sub-zero cold under two metres of ice.

This probable USO (Unidentified Submarine Object) was one example among many similar reports over the centuries when huge spaceships have been observed to enter, then rise from the sea.

Mariners of old spoke of 'sea serpents'. Were some of these creatures USOs?

A UFO or USO which hovered over the Wanaque Reservoir in Northern New Jersey on 12th January, 1966, left an 18-m hole through 5 cm of ice. Another unaccountable fact was that early in 1968 an 'incredibly powerful thing' made two immense holes through 90 cm of ice in two Central Sweden lakes near Malung and Serna. Frogmen could find nothing, but it was considered that *something* had gone through the ice!

WAS CORN MOTHER ONE OF THE 'SKY PEOPLE'?

Indian legends from Canada, and legends from the Buffalo Hunters of the Great Plains, relate the story of 'Corn Mother', a lady from the skies who brought the gift of maize from the heavens. Corn Mother taught many things about the heavens; about the planets, stars, the Moon, and Sun. She also spoke of the 'Sky People'.

The Great Plains Indians believed that tobacco was also brought from the sky; it was known as a 'Star-plant'. Of interest in this connection is a medicinal plant named 'Star-reed' which grows in Peru, and a substance 'Star-shoot' found in marshes, which was believed to have arrived on Earth from a shooting star. Were all these items 'gifts from the gods'?

Corn Mother bears similarity to Isis in Egyptian legends. Isis was reputed to have brought the gifts of red and white barley, wheat, and corn, from the heavens; but American Indians could not have known of Egyptian legends. America was undiscovered in those days or was it? Remember Buddha's footprints in Idaho?

Two Leningrad specialists of Oriental antiquity discovered that ancient Asian geographers knew of the Americas at least 1,500 years before Christ. The Russian specialists, presenting their report in 1970, said they based their conclusions on the deciphering of ancient Tibetan maps, previously thought to be portraying 'imaginary lands of fantasy or mystical Buddhist tales'.

DID THE ACOMA, HOPI, AND ZUÑI, LEARN BOOMERANG SECRETS FROM SPACEMEN?

The Acoma, Hopi, and Zuñi tribes of the American South-west have used the boomerang since antiquity. They say the secret of boomerang construction was taught to them by the gods, who were careful, however, to warn them of its danger, and to use it only for hunting.

The three tribes also relate that the gods taught them how to construct four other types of weapons; the bow and arrow were probably one other type.

All the following lands possessed the boomerang in antiquity: Australia, the only returning boomerang; Borneo, Celebes, Ethiopia, Egypt, India, and France. Did cosmic beings once visit these ancient lands and teach boomerang construction to the inhabitants?

Many prehistoric Red Indian symbols were discovered on 'Newspaper Rock' at Indian Creek, Colorado/Utah. Two symbols, among others, were circular objects with central sections, and emitting rays. Possibly UFOs in which the Red Indians' deities arrived?

THE FLYING BASKET WITH TWELVE STAR MAIDENS

Wampee, a 'great hunter' of old, recorded his tale of the sky 'basket'. Red Indian legends state that Wampee once watched a 'flying basket' (UFO?) containing twelve 'star maidens' descend to the ground. He managed, through guile, to capture one of the 'star maidens' whom he married. But as she was 'a daughter of one of the stars', she grew very homesick. Through secret efforts, the star maiden managed to obtain another 'basket' and once more ascend to her home in the stars.

HIAWATHA 'ROSE TO HEAVEN' (IN A UFO?)

Was Haio-hwa-tha or Hiawatha, a spaceman? It was said that:

> 'Hiawatha rose to heaven in the presence of the multitude, and vanished from sight in the presence of sweet music.'

Hiawatha was a chief of the Mohawk; was he taken off in a UFO? The 'sweet music' mentioned in the above quote is similar to a temple record description of a Hindu Vimana, at Ajodhyá in North India. The temple record speaks of dulcet tones emitted by the Vimana as it coursed through the firmament on its airy way.

Ajodhyá, on the Gogra River bank, was the archaic capital of Oudh or Oude near Fyzabad or Faizabad.

DO LEMURIANS, MUANS OR ATLANTEANS LIVE UNDER MT. SHASTA?

Mt. Shasta is an extinct volcano in the Sierra Nevada, California. Inexplicable incidents relating to Mt. Shasta have been recorded for a long time; it is reputed to have a subterranean home, the abode of a society of secret beings.

An unaccountable light of strange luminosity has been seen to illuminate large areas of the Mt. Shasta slopes at times; this is considered by those who have seen it to be undoubtedly of artificial origin. On occasions, the Mt. Shasta beings themselves are reputed to have been seen; they are said to resemble the ancients with long toga-like robes.

The Mt. Shasta beings are believed to have donated gifts to the American Red Cross during the 1914–18 war, and during the nineteenth century to have purchased goods in the local towns with gold nuggets.

A Dweller on Two Planets, published in 1884 and written by a local resident Fredrick Oliver, describes the Mt. Shasta beings. It seems that Oliver may have been invited to a possible underground home of the beings because he wrote with such detail.

According to Oliver, these beings are surviving Atlanteans; they possess magnetic-powered spacecraft, and are in contact with Venusians. UFOs have been sighted in this area! In October 1956, one UFO was seen to leave a formation of fourteen, and descend on top of Mt. Shasta!

Oliver said that these surviving (Atlanteans?) once owned aerial vehicles with wings, and powered by gas-reservoirs, but these were superseded by craft flying by magnetic forces.

This nineteenth-century book states that the Mt. Shasta people could produce fireballs of different colours, namely: red, orange, yellow, and green. In 1951, nine green fireballs were seen together in the skies over the Albuquerque, New Mexico vicinity; they were described as 'bright as the moons'. Many more of these strange green fireballs were seen in 1951, but *only* in the southwest where the Los Alamos atomic energy plant is located.

The fireballs had the appearance of burning copper or a green neon tube; on the spectrum chart the green colour was 5,200 angstroms which was matched to burning copper. Meteorites do not contain copper – any copper in meteorites is oxidized as the meteorites enter our atmosphere – but *heavy concentrations* of copper particles were found in the New Mexico and Arizona air after the green fireballs had vanished!

Why were the green fireballs seen only in the southwest, and particularly over the Los Alamos atomic energy installation? Where was their point of origin?

Wherever they came from, and whoever sent them, one idea was that their purpose was to mop up excess radiation after atomic tests!

Oliver's description of the Mt. Shasta dwellers' exotic subterranean abode is similar to a story in *Doctor From Lhasa* by Lobsang Rampa, who claims to be a reincarnated Tibetan monk. Rampa described a visit to a secret cave under the Potala palace in the Tibetan capital Lhasa. While he was training to become a doctor-priest his mentor led him to one of the caverns and showed him the work of an ancient race. Records were exquisitely engraved on a spacious golden panel, while in addition there was an antiquated star atlas. But due to its hoary age all stars were in different positions.

SARCEE AND OJIBWA INDIANS OSTENSIBLY JOURNEYED TO VENUS AND URSA MAJOR

Two old legends from Canada, although poetically written, contain quite engaging information relating to 'star-people', and other curiosities associated with beings who reside in the heavens. The two legends condensed and outlined here, are examples of other similar tales from Canada of old. Are these legends true?

The first legend, from the Sarcee people, relates the

story of a young man who travelled to Earth from the Evening Star, i.e. Hesperus or Venus. The sky man, bewitched with admiration for the charms of a winsome Indian girl, married her, and together they travelled to the land of the 'star-people'. After a time a son was born, but when the small boy was some three years old, the Indian girl, his mother, grew homesick for Earth. The story finally recounts that she beseeched the 'star-people' who brought her back to Earth in a 'flying object' (unidentified?), and departed for the heavens.

The second story of Fisher Maid, from the Ojibwa Indians, is even more enchanting and unwonted.

The tale is of the birth of a baby girl who was named Otchiganse-Equa. At quite an early age it was observed that she was a most unusual child, not at all like other youngsters. As the years passed, Otchiganse-Equa developed marvellous and distinctive traits which were wholly dissimilar to those of other people. One fine day a 'Star-Man' descended from the heavens and married her to the accompaniment of many sky beings singing divinely at the wedding.

The narrative further recites that eight sky women of great beauty arrived on Earth and abducted Otchiganse-Equa's husband Otchiganang. In great consternation Otchiganse-Equa changed into a 'dazzling white fisher' (or entered a UFO?) then flew to the land of the sky people and killed the eight women. This in turn inflamed the sky beings so much that they chased her through the heavens hurling bolts of 'lightning' (or A-missiles?) at her.

Eventually, Otchiganse-Equa was reunited with Otchiganang and they travelled to dwell peacefully on a 'rock' named 'Akakoojish' orbiting around 'Fisher-Star' in the Great Bear.

Is Akakoojish a planet of a solar system in the Ursa Major or Great Bear constellation, and is 'Fisher-Star' their sun?

Otchiganse-Equa was quite unlike ordinary mortals when born; the legend inferred that she was a space being reborn on Earth.

DIVERS AND DIVERSE PUZZLES FROM THE AMERICAS

The 'Jurupari' religion of Brazil among the Arawaks and Tupi native tribes appears to be closely related to Freemasonry. Freemasonry is thought to have originated in the Biblical lands or Egypt, but it is also suspected that the Mayas and Quichés of 11,500 years ago had a possible connection (shortly after Atlantis sank?).

Did Freemasonry have an Atlantean origin? How was it disseminated to different areas of the globe?

PERHAPS SOME TEOTIHUACÁNOS LEFT MEXICO BY SPACECRAFT?

The original inhabitants of North Africa between the Mediterranean and the Sahara, the Berbers, whom it is suspected by some to have migrated from Atlantis, relate in their legends that large numbers of Earth folk left, or were abducted, from Earth by spaceships in antiquity.

In an area ranging from Chile to Southern Mexico, civilization appeared very suddenly between 3000 to 1000 B.C., and in some areas, disappeared just as rapidly and mysteriously a few hundred years later. Where did (all) the people go? And from whence did they come?

The Olmec or La Venta culture was flourishing in the Mexican jungles – the Chiapas, Veracruz and Tabasco rain forests – then they vanished about A.D. 600. Theosophists believe there was an ancient link between Olmecs and Atlanteans!

On one of the exceedingly heavy stone heads left by this civilization, is a strange headdress with *Ethiopian* features *quite unlike* the features of ancient Mexican peoples! The headdress has been described as 'like the helmet of a football player' and 'like a boxer's training gear'. A space helmet representation?

The basalt heads are far too heavy to move, but the Olmecs transported the rock from a quarry 160.9 km away; by anti-gravity or spaceships?

On a site 51 km 428 m north of Mexico City in the San Juan – Teotihuacán zone, archaeologists are attempting to discover why the population of another ancient civilization suddenly disappeared without reason, somewhere between A.D. 800 and 1116.

Teotihuacán – 'city of the gods' – was partly excavated by Spanish Conquistadores in the sixteenth century; they found huge human bones which they sent back to Charles V (1500–58), then king of Spain. The unusually large bones confirmed both Toltec and Aztec legends of a race of giants once living at Teotihuacán.

Why did the population abruptly vanish from Teotihuacán? Were some Teotihuacános taken off by the giants in spaceships? Were some Teotihuacános themselves ephemeral or transitory space beings dwelling on Earth for only a transient period before moving on to distant spheres – or back home?

The ruins of Angkor Wat in the Cambodian jungle testify to yet another civilization who deserted their city. the ninth-to-fifteenth-centuries Indo-European Khmer Empire people of Angkor Wat left carvings of *winged deities* in their temples!

In *The Eagle – Spacecraft of the Pre-Scientific Age* by José Patrocinio de Souza (*Unesco Courier*, June 1970) the author notes the large number of times that the Eagle is portrayed in ancient art works carrying off humans into space. Did the Eagle symbolize a spaceship?

THE EAGLE KNIGHTS BASED THEIR UNIFORM AND HEADGEAR STYLES ON A SPACE SUIT AND HELMET!

The Aztec warrior knights of the 'Eagle' Regiment adopted a most curious uniform. This was covered with Eagle feathers, and was pulled-in at the wrists and just below the knees. The headdress of the Eagle warrior was in the shape of an Eagle's head covered with Eagle feathers, and with a large open beak. When dressed in full uniform with the headdress covering the warrior's head, his appearance took on an interesting aspect, which resembled a being wearing a space suit and helmet!

Did this warrior regiment adopt their uniform and

headdress in memory of a spaceman god, or spacemen gods, when he or they appeared to the Aztecs wearing a space suit and helmet, or space suits and helmets – possibly even centuries earlier?

The Eagle Knights were fanatical believers in their gods – even to the point of performing human sacrifices to them. These gods, or spacemen, who evidently came in peace to assist and teach ancient Earth races, must have experienced bitter disappointment if they later learned that human sacrifices were performed in their honour after they had left Earth!

A photograph of an Aztec carved stone head, probably portraying the fourteenth-century Aztec god Huitzilopochtli or 'hummingbird-on-the-left', has facial features which seem not of this world, being advanced, superior, or even with a possible hint of arrogance, combined with majesty, bearing, and wisdom. The facial features of this stone carving cannot be related to Aztec features, which had a different nose, mouth and cheekbone structure.

It seems unlikely that so much time and skill would go into a carving of a warrior of the Eagle regiment; this particular sculpture seems undoubtedly to be of a *special* entity – *a god from the skies*!

The helmet on this carving is, apparently, a *replica of a space helmet!* Archaeologists would, of course, identify it as an eagle's head, but what self-respecting eagle has an extension (at the base of the helmet) which so closely resembles six tube-like appendages – respiratory pipes/electronic cables – leading into a space helmet? Where the helmet is bent in a V-shape must be where a pivot on each side moved the face visor on the real space helmet. The whole appearance of the headgear is very similar to a modern space helmet!

This (slightly damaged) carving much more closely resembles a space helmet than any eagle's head or snake's head. Even similar copies of this helmet worn by Aztec gladiators *must* have been copies of the space helmet worn by an earlier god, or gods. The snake's head, or eagle's head were probably the nearest resemblances ancient Mexican races could find to identify with the space helmets worn by their spacemen gods. Precisely the same parallel is found among other ancient races, e.g. the Sumerians related a god (Oannes the 'man-fish') to a creature common to that area – a god with a human head enclosed within the head of a fish, again probably a space helmet!

The symbolic snake's head among ancient Mexican races may have an alternative meaning, perhaps even additional meaning, when we remember the 'serpent' space beings!

A SPACE HELMET DRAWN ON A MAYAN CODEX?

Only three Mayan manuscript fragments escaped burning by the Spanish Conquistadore Diego de Landa in 1562. As already mentioned, they are kept in the museums of Dresden, Paris and Madrid. The Motul and Chilam Balam are the only guides to deciphering the ancient manuscript fragments. The Motul, consisting of 35,000 words, is a Mayan dictionary compiled by a missionary in the colonial period, while the Chilam Balam of 64,000 words is a book of ancient Mayan prophecies recorded in Latin from archaic Mayan legends during the same period.

After processing 99,000 words from the Motul and Chilam Balam by electronic computer, scientists at the Institute of Mathematics in Novosibirsk succeeded in translating sections of the extant Mayan works. One translated sentence refers to a maize god who excelled in fashioning pottery from white clay.

On one of these extant manuscript fragments are drawn several unusual heads; one has most unusual facial markings, while an even more interesting *glowing* globular protrusion rests on the cranium. The markings on the glowing protrusion resemble the markings and emanating rays on, and surrounding the heads of, spacemen rock drawings in various parts of the world.

The Mayas were exceptionally skilled in astronomy, mathematics, teeth inlaying, and skull surgery; from whom did they learn?

An early link may also have once existed between the Mayas and Atlantis, in view of the architectural style of the Andros temple ruins, which in lay-out is identical to the 'floor plan' of the Mayan 'temple of the turtles' at Uxmal, Yucatán. Did the Mayas come from Atlantis?

MANCO CCAPAC SUDDENLY APPEARED WEARING GOLDEN ARMOUR

The Bridge of Light by US archaeologist and author, Professor A. Hyatt Verrill, describes a 'bridge of light' over a dangerous gorge, the only access to a certain pre-Inca city. Professor Verrill, who specialized for thirty years in studying ancient and lost civilizations in south and central America, claimed that the stories related in his book were 'much more than a legend'.

According to certain extant legends the Incas apparently arrived on Earth from another planet. The 'first Inca' was Ayar Manco Ccapac who about A.D. 1000 suddenly and unexpectedly appeared in Peru wearing a handsome suit of 'golden armour' – or a space suit?

The Incas, like the Hindu deities Sūrya-Vansa, were a 'solar race', or as the Incas preferred to call themselves 'Sons of the Sun' (or 'Paccari-Tampu' – the place of our origin).

It seems likely that the Incas descended from Ccapac, three other spacemen and four spacewomen, all special deities, plus the two tribes Maras and Tambo, all of which had a 'sudden' origin, and were associated with the Sun-God!

Machu Picchu, the famous legendary 'lost city' of the Incas, discovered near Cuzco in the Peruvian Andes in 1911 by the US historian Hiram Bingham, has stonework fitted together so precisely, that a razor blade cannot pass between the blocks. It is also still a mystery how the Incas carried huge boulders up the mountain without wheel or vehicle. By anti-gravity?

One day Machu Picchu was deserted, and around this same time, a mystery civilization appeared in the Great Rhodesian Veld in Africa and built similar constructions, the most famous of which is Great Zimbabwe (Zim-ba'-bwe). Heated controversy surrounds the theoretical identity of the people who built Zimbabwe, but could Zimbabwe have been the still undiscovered 'Vilcabamba', the legendary 'last refuge' of the Inca civilization to escape the Spanish Conquistadores in 1533? This is not to be confused with Vilcabamba village, an isolated community of early Spanish extraction in the Ecuadorian Andes.

Zimbabwe, an ancient fortress town of vast proportions, was constructed between the eleventh and fifteenth centuries A.D., or possibly as late as the seventeenth century. The antecedent estimated construction date of A.D. 200–300–400 is now considered inaccurate.

It has apparently been established that (Great) Zimbabwe was built by a race of advanced aliens, not Bantus, but who were they? The ruins of Zimbabwe, south-east of Fort Victoria, were discovered in 1868 by the ivory hunter Adam Renders, but the builders of this intricate complex have not been identified. No African race today constructs in like fashion, while the gold artifacts, pottery, and sculptures unearthed in the area indicate the skill of an alien people. The technique for smelting gold also indicates the work of a non-indigenous race. Two Zimbabwe facts are prominent. In the 'Great Temple' the granite blocks, like the constructions at Machu Picchu, were fitted together so closely and accurately that mortar was not required, and secondly, the Temple's eastern wall extends 79.5 m, which is the exact area covered by the sun's rays at midsummer or solstice. Pointedly, the inhabitants of Machu Picchu calculated the solar year from the shadow cast by the giant carved rock 'Intihuatana' or 'stone of the Sun'.

Archaeologists cannot establish with certainty who the builders of Zimbabwe *really* were. Accusations of 'racists and white supremacists' have been levelled at archaeologists who have proposed that Zimbabwe was not the work of an indigenous African ethnic group, but the fact remains that the architectural style of Zimbabwe has noticeable similarities with architecture in Gôa, the

Indus Valley, Johore on the Malay Peninsula, Persia, and South Arabia, but cannot be identified with any one place; while suggested builders have been Carthaginians, Phoenicians, Romans, Sabaeans, a Semitic people who inhabited south-west Arabia, and Egyptians. Stone birds resembling the Egyptian Horus depiction have been excavated on the Zimbabwe site.

Several similarities are evident between Zimbabwe and Machu Picchu architecture, while the mystery is further compounded by the disappearance of the population from Machu Picchu and the appearance about the same time of an unidentified civilization in the Great Rhodesian Veld.

Did the inhabitants of Machu Picchu travel to the Great Rhodesian Veld? If so, how could the entire population of a city in the Andes all migrate (suddenly?) to another country? By spaceships? Rhodesia is in a direct line from Peru, parallel with the Equator, across Brazil and the Atlantic Ocean!

THE PEOPLE WHO ARE 'OLDER THAN THE SUN'

The beautiful Lake Titicaca is high in the Andes between Bolivia and Peru. The great mystery associated with the lake is that it seems to have once been a sea harbour; the remains of a well-defined seashore line can be seen, while sea weeds and shells have been found there. When, and how, did the harbour rise nearly 3·9 km to become a mountain lake? Perhaps the sinking of Lemuria or Mu explains the Titicaca enigma?

Another observation of equal fascination is the 'Uru', an Indian tribe now numbering less than 200 who live on the banks of Lake Titicaca. They are the oldest and the original inhabitants in the area. Is there a possible remote link to the Maori's 'Uru', the original homeland according to one legend?

When asked of their history the Uru of Lake Titicaca say their tribe is 'older than the Sun'. Does this mean that their ancestors once lived in a *different* solar system before our own star system came into existence?

12. KING ARTHUR ARRIVED IN A 'WINGED DRAGON'?

'... It seemed in heaven, a ship, the shape thereof a dragon wing'd, and all from stem to stern Bright with a shining people on the decks ...'

(Alfred Lord Tennyson 1809–92.
From: *The Coming Of Arthur*.)

The volume of material about King Arthur recorded since the sixth century A.D. cannot be disregarded. Throughout Britain, Europe and notably Italy, can be found an abundance of data.

One of the oldest works relating to King Arthur is *The Acts Of The Illustrious King Arthur*, written by Gildas Badonicus (A.D. 516–70). It was around this same time that Arthur and Guinevere were said to have lived.

Nennius the Historian, who flourished in A.D. 796, was author of *Historia Britonum* or *History of Britain*. Nennius wrote the most valuable stories from the point of view of King Arthur as an historical fact.

One early legend associated Arthur with the Great Bear or Ursa Major constellation; what could this mean? King Arthur was believed to be the son of Uther-Pen-Dragon (King of Britain or Wales) who received the name of 'Pen-dragon' after witnessing the startling sight of a fiery flying 'Dragon'.

A legend of mediaeval times relates that Arthur was brought to England as an infant in a 'Winged Dragon'. Was Uther really the father, or did he witness the coming of the 'Winged Dragon' with the baby Arthur? In twentieth-century terminology, was Arthur brought to England in a spaceship? Was he a spaceman?

In the year 1139 Galfredi Monumentensis or Geoffrey of Monmouth, wrote his book *Geoffrey's British History*. In this work is the translation of an old Breton-Celtic

(Brezonec) manuscript found in Brittany, and dated at either A.D. 497 or 508.

The following are excerpts from three chapters of this ancient chronicle, which refer to the arrival of the flying 'Dragon'.

CHAPTER XIV

'During the transactions at Winchester, there appeared a star of wonderful magnitude and brightness, darting forth a ray, at the end of which was a globe of fire in form of a dragon, out of whose mouth issued forth two rays; one of which seemed to stretch out itself beyond the extent of Gaul,* and the other towards the Irish sea, and ended in seven lesser rays.'

CHAPTER XV

'At the appearance of this star, a general fear and amazement seized the people, and even Uther, the King's brother, who was then upon his march with his army into Cambria,† being not a little terrified of it, was very curious to know of the learned men, what it portended. Among others, he ordered Merlin to be called, who also attended in this expedition to give his advice in the management of the war; and who, being now presented before him, was commanded to discover the significance of the star.'

CHAPTER XVII

'And (Uther) remembering the explanation which Merlin had made of the star above mentioned, he commanded two dragons to be made of gold, in likeness of the dragon which he had seen at the ray of the star. As soon as they were finished, which was done with wonderful nicety of workmanship, he made a present of one to the Cathedral Church of Winchester, but reserved the other for himself, to be carried along with him to his wars. From this time,

*France.

†Wales.

therefore, he was called Uther Pendragon, which in the English tongue signifies the dragon's head; the occasion of this appellation being Merlin's predicting, from the appearance of the dragon, that he should be king.'

ANALYSIS

The Chapter XIV excerpt states that the very bright 'star' sent forth 'a ray, at the end of which was a globe of fire in form of a dragon'. In modern terminology the 'star' could have been a huge UFO carrier or 'mother ship'; and the 'ray' sent forth with a 'globe of fire' could have been a UFO, or possibly several UFOs, being released. *Is this how King Arthur really arrived in England?*

Arthur was believed to possess two castles; one was at Tintagel in north Cornwall, where legend speaks of the arrival of the flying 'Dragon' craft. The ruins of Tintagel Castle still remain and the castle has been proved to have been constructed about the sixth century A.D. At the foot of the cliff under the castle, is a cave reputed to have once belonged to Merlin.

Arthur's chief castle at Camelot was undiscovered, but in an attempt to locate the site, archaeologists processed archaeological data by electronic computer. Subsequently, excavations around South Cadbury Hill, Somerset, disclosed a section of wall, and fragments of wine jars dating back to the fifth or sixth century A.D.

Possibly significant is that the Cadbury Hill excavations proved that the military defences (of the remains) had been strengthened about the sixth century A.D. Tintagel Castle was *also* strengthened for military use about the same time, probably shortly after construction!

UFOs STARTLED THE POPULATION OF MEDIAEVAL ENGLAND

Since the start of the mediaeval era (A.D. 476), many strange aerial incidents were recorded. One of the oldest records from this period is The Venerable Bede's

Ecclesiastical History of England. Bede (A.D. 673–735) was an English priest and historian, and the following two excerpts from the *Ecclesiastical History* relate to unusual aerial events sighted from certain monasteries:

A.D. 676

'... on a sudden, a light from heaven like a great sheet came down upon them all and struck them with so much terror that they in consternation left off singing. But that resplendent light which seemed to exceed the sun at noonday, soon after rising from that plane, removed to the south side of the monastery, that is to the westward of the oratory, and having continued there some time and covered those parts in the sight of them all, withdrew itself up again to heaven ... this light was so great that one of the eldest of the brothers who at the same time was in the oratory with another younger than himself, related in the morning that the rays of light which came in at the crannies of the doors and windows, seemed to exceed the utmost brightness of daylight itself.'

This sighting could not be a natural phenomenon for two reasons; the first is that after descending, the 'light' rose, then moved to another area outside the monastery, and the second reason is that after staying long enough for all the brothers to see, the 'light' 'withdrew itself up again to heaven'. This proves that the 'light' was *controlled*, and in all probability this is a genuine UFO report from the Middle Ages!

A.D. 729

'In the year of our Lord's incarnation 729, two comets appeared about the sun, to the great terror of the beholders. One of them went before the rising sun in the morning, the other followed him when he set at night ... They appeared in January, and continued nearly a fortnight.'

ANGLO-SAXON CHRONICLERS RECORDED MANY UFO SIGHTINGS

The *Anglo-Saxon Chronicle* contains many Middle Ages UFO sightings of great interest; for example:

A.D. 678

'This year the star called a comet appeared in August, and shone like a sunbeam every morning for three months.'

A.D. 776

'In this year a red cross appeared in the sky after sunset.'

Now this is most interesting. The Roman Emperor Constantine The Great (A.D. 288–337) was converted to Christianity after seeing a cross in the sky in the year 312. Aethelwerd's *Chronicle* for the year 773 states: 'The sign of our Lord's Cross appeared in the heavens after sunset.'

Several other historical occasions when flying crosses were seen, were in Poland in 1269; Switzerland 1478; Holland 1528; Italy 1547; Germany 1554 and 1561.

In September 1967, flying crosses were again seen over England, in the Southern Counties. An official attempt to discredit the sightings by saying they were US planes re-fuelling in midair failed to convince many people. Could US planes have been re-fuelling in mid-air in A.D. 312–1561?

A.D. 793

'In this year dire portents appeared over Northumbria and sorely frightened the people. They consisted of immense whirlwinds and flashes of lightning, and fiery dragons were seen flying in the air.'

'Fiery dragons' or UFOs?

A.D. 1097

'Then at Michaelmas, on the fourth before the Nones of October, an uncommon star appeared shining in the evening, and soon going down; it was seen in the south-west, and the light which strove from it seemed very long, shining towards the south-east; and it appeared after this manner nearly all the week.'

A.D. 1104

'This year, the first day of Pentecost was on the Nones of June, and on the Tuesday after, at midday, there appeared four circles of a white colour round the sun, one under the other as if they had been painted. All who saw it wondered, because they had never remembered such before . . .'

Four 'white' 'circles' 'one under the other as if they had been painted'. This daytime sighting definitely indicates four UFOs in formation!

A.D. 1105

'On the eve of Cena Domini, that is the Thursday before Easter, two moons were seen in the sky before day, one to the east and one to the west.'

A.D. 1106

'In the first week of Lent, on the evening of Friday the fourteenth before the Kalends of March, a strange star appeared, and it was seen awhile every evening for a long time afterwards. This star appeared in the south-west; it seemed small and dim, but the light that stood from it was very bright, and like an exceedingly long beam shining to the north-east; and one evening it seemed as if a beam from over against the star darted directly into it. Some persons said that they observed more unknown stars at this time, but we do not write this as a certainty because we saw them not ourselves.'

This report mentions one, and possibly several, unidentified 'stars' not behaving as stars should. Was the 'beam' that 'darted directly' into the 'star' a UFO returning to the 'Mother Ship'?

Isaiah, whose work was dated from 760–698 B.C., recorded a similar phenomenon in Biblical phraseology:

'Who are these that fly as a cloud, and as the doves to their windows?'

(*60:8*) 698 B.C.

Did Isaiah also see UFOs returning to the carrier ship?

A.D. 1110

'. . . After this in the month of June there appeared a star in the north-east and its light stood before it to the south-west, and it was seen for many nights, and ever as the night advanced, it mounted upwards and was seen going off to the north-west.'

Now, have you ever seen a 'star' mount 'upwards' and go off to the north-west? Is more convincing proof of an ancient UFO report necessary?

A.D. 1114

'In the end of May this year, a strange star with a long light was seen for many nights.'

SHAKESPEARE PROBABLY SAW FIVE UFOs ABOUT 1596

The most curious words that Shakespeare made Hubert say in the play *King John* (1596) were:

'My Lord, they say five moons were seen to-night;
Four fixed, and the fifth did whirl about the
Other four in wonderous motion.'

STONEHENGE MAY BE BOTH AN ASTRONOMICAL COMPUTER AND THE CHIEF LINK IN A QUARTZ CRYSTAL ULTRA-VIOLET RAY ENERGY NETWORK

Who built Stonehenge?* It is now known *why* it was constructed. *Two* vital discoveries have been made. The

*circa 1800–1400 B.C.

distinguished scientist Dr. Gerald S. Hawkins proved by electronic computer that Stonehenge was once utilized by the ancients as an astronomical computer of precise accuracy.

The second important discovery is that Stonehenge is a *link* in a gigantic 'power network'.

John Williams, a solicitor of Abergavenny, Monmouthshire, studied the relationship of more than 3,000 standing stones and stone circles on Ordnance Survey maps. His calculations proved that all the stone circles and standing stones are aligned to each other at up to a 32 km 180 m distance, with a 23½°* angle, or a multiple of 23½°.

Over a period of years Williams photographed thousands of the standing stones; many of his photos when developed had a fogged band across them. A colour photo had a dark blue fog band. Yet Williams' camera and film were in perfect condition, and a friend using a different camera and film obtained the same fog results. Most of the standing stones, if not all, were discovered to contain quartz and to be emitting *ultra-violet rays*! (Quartz crystal was utilized by the 'serpent' beings!)

Williams researched his discoveries since about 1951 and found that over 200 of the sites are aligned in a north–south arrangement. He believes that if the stone circles and standing stones are segments in a gigantic 'power network' then the 'roofing stones' on top of Stonehenge were once used as 'rocking stones' to start the 'power network' operating.

Is it really possible that ancient Britons knew how to construct a highly advanced system such as this?

The discovery that Stonehenge was once used as an astronomical computer is a major breakthrough, but the disclosure that it had a dual purpose – a link in a mystery power network as well – is a staggering thought! The site for Stonehenge must have been calculated by advanced mathematics for its dual purpose. It *must* have been the work of advanced and superior beings, but who were they? Druids, Atlanteans, cosmic entities?

*Curiously, 23½° is also the axis tilt of Earth.

What of the huge stones, and in particular the 'roofing stones'? It seems a formidable and overwhelming task to have raised these stones by manual labour; were they lifted by anti-gravity, or by slings under spaceships? And why did the builders mysteriously transport the 'bluestones' 240 km from Pembrokeshire when large masses of rock are found only 32 km from Stonehenge? Did the bluestones possess particular ultra-violet properties? If so, what was, or is, the purpose of the 'power network'? Does it bear any relationship to UFO power systems? Let us see . . .

CO-ORDINATED SCIENTIFIC ACTIVITIES UNDERTAKEN BY SPACEMEN IN 1646

Weird and mysterious things happened in England on 21st May, 1646. 'Ships in the Ayre' were sighted, a 'speare' descended from 'Heaven', and 'Men in the Ayre' were also seen.

The following description is from a book in the British Museum Library:

> 'Signes from Heaven: or SEVERALL APPARITIONS seen and heard in the Ayre, in the Counties of Cambridge and Norfolke on the 21 day of May last past in the afternoon, 1646.
>
> ---
>
> Viz
> A Navie or Fleet of Ships under Sayle.
> A Ball of wilde-fire rolling up and downe.
> Three men struggling with one another, one having a Sword in his hand.
> Great hailstones round and hollow like Rings.
> Extraordinary beating of drums in the ayre, &c.
> A Pillar or Cloud ascending from the earth like a spire-Steeple, being opposed by a Speare or Lance downward,
> Being made manifest by divers and several Letters from persons of credit in both Counties, and sent up to his City to their friends for Truth.
>
> ---

LONDON

Printed by T. Forcet, dwelling in Old Fish-street, in Heydon-court, 1646.

STRANGE Signes from Heaven, to warne and awaken the Eastern Association, with the Southern parts of the Kingdome.

XXX Pon the one and Twentieth day of May in the afternoone,

XVX in the year of 1646, there were very strange sights seen,

XXX and unwonted sounds heard in the Ayre, in severall places as followeth.

About New-Market in the county of Cambridge, there were seen by divers honest, sober and civill persons, and men of good credit, three men in the Ayre striving, struggling, and tugging together, one of them having a drawn Sword in his hand, from which Judgement God in mercy preserve these three Kingdomes, of England, Scotland, and Ireland, from further conflicts and effusion of blood.

Betwixt Newmarket and Thetford in the aforesaid County of Suffolk, there was observed a pillar or a Cloud to ascend from the earth, with the bright hilts of a sword towards the bottom of it, which pillar did ascend in a pyramidall form, and fashioned itself into the form of a spire or broach Steeple, and there descended also out of the skye the form of a Pike or Lance, with a very accute point out of the skye likewise, which was ready to interpose, but did not engage it self.

The first Speare which came down from Heaven point blanck, was after a while clean elevated higher, and that spire or Speare which went up from the earth, ascended after it, to encounter with it a second time.

This continued about an houre and a halfe.

At Sopham in the County of Cambridge aforesaid,

a ball of wilde-fire fell upon the earth, which burnt up and spoyled about an Aker of Graine, and when it had rolled and runne up and down to the terror of many people and some Townesmen that see it, it dissolved and left a most sulpherous stinck behind it.

Also at Camberton in the County aforesaid, divers of the Trayned Bands being met at a Muster, did behold the forme of a spire Steeple in the Skye, with divers Swords set round about it.

Also at Brandon in the County of Norfolke the inhabitants were forced to come out of their houses to behold a Pike or lance descending downward from Heaven. The Lord in mercy bless and preserve his Church, and settle Peace and truth among our Church men.

In Brandon in the County aforesaid, was seen at the same time a Navie or Fleet of Ships in the ayre, swiftly passing under Sayle, with Flags and streamers hanged out, as if they were ready to give encounter.

In Marshland in the County of Norfolke aforesaid, within three miles of Kings Linne, a Captain, and a Lieutenant, with divers other persons of credit, did heare in the time of Thunder, a sound as of a whole Regiment of Drums beating a call with perfect notes and stops, much admired by all that heard it.

And the like Military sound was heard in Suffolk upon the same day, and in other parts of the Eastern Association.

In all these places there was great Thunder, with Raine and Hailestones of extraordinary bigness and round, and some hollow like rings.

The Lord grant that all the people of this Kingdome may take heed to every warning Trumpet of his, that we may speedily awaken out of our sins, and truly turn to the Lord, fight his battles against our spiritual Enemies, and get those inward riches of which we cannot be plundered of, and so seek an inward Kingdome of Righteousness and Peace, that we may be capable in his good time of a settled peace, and stare in the outward Kingdome, and all through our Lord Jesus Christ.'

FINIS

What really happened on that day in 1646? Were spacemen landing and conducting aerial manoeuvres in UFOs? The local populace apparently thought that 'Signes from Heaven' were warning them to repent their sinful ways, but perhaps this old UFO report could be interpreted in a *different* way. Were spacemen undertaking *scientific activities* on 21st May, 1646, and *not* just performing aerial displays for the astounded population?

Discoveries by John Williams may have a definite and significant bearing on the 1646 enigma. Were UFOs conducting scientific activities relating to a mysterious power system on that afternoon in seventeenth-century England? On occasions, UFOs have been observed in the midst of an unusual operation; i.e. they hover over high-tension power cables then extend a slender rod until it contacts the power line. After a brief period, the UFO withdraws the extension and moves off, giving the impression to eye-witnesses that it has been recharged by electrical energy.

Williams discovered that Stonehenge is a link in a gigantic 'power network', and up to a 32 km 180 m distance exists between the stone circles and standing stones in nearby sites. Let us now break this discovery down into component parts and study each one. The areas mentioned in the 1646 report are Norfolk, approximately 257 km 440 m from Stonehenge to the centre of Norfolk, an area between Newmarket in Suffolk and in Thetford in Norfolk about 217 km 215 m from Stonehenge; and Newmarket, about 193 km 80 m from Stonehenge. Very approximately, a 32 km 180 m distance separates these sites.

The 'Signes from Heaven' report states that 'Betwixt

Newmarket and Thetford' 'there was observed a pillar or a Cloud to ascend from the earth, with the bright hilts of a sword towards the bottom of it, which pillar did ascend in a pyramidall form, and fashioned itself into the form of a spire or broach Steeple,* and there descended also out of the skye the form of a Pike or Lance, with a very accute point out of the skye likewise, which was ready to interpose, but did not engage it self'.

Was the 'pillar' or 'Cloud' ascending 'from the earth' an electrical discharge, LASER beam, or ionized gas forming a column relating to a recharging operation, or (to us) an unknown scientific operation? The 'Pike' or 'Lance' 'with a very accute point' descended from the 'Skye' and attempted to meet with the 'pillar' or 'Cloud'. The 'Speare from Heaven' was apparently drawing the 'pillar' or 'Cloud' higher into the 'Ayre' without actually touching it, as if a magnet was drawing an iron bar towards itself. The 'pillar did ascend in a pyramidall form, and fashioned itself into the form of a spire or broach Steeple . . .' Was the pyramidal shape an optical illusion due to the 'pillar' or 'Cloud' ascending to a great height, similar to a long road also assuming an elongated pyramidal appearance as it recedes into the distance?

The report also mentions that in Newmarket there were seen: 'Three men in the Ayre striving, struggling, and tugging together, one of them having a drawn Sword in his hand . . .' Were spacemen on a UFO deck, or suspended alongside a UFO with anti-gravity devices or rocket belts, attempting to line-up a UFO extension (or 'Sword') with a discharge (artificial lighting, LASER beam or plasma?) from the ground? There seems to have been difficulty in lining-up the extension ('Sword') without manual assistance, as the report from 'Betwixt Newmarket and Thetford' gives the impression that the 'Speare from Heaven' had several attempts to meet with the 'pillar' or 'Cloud' over a period of 'an houre and a halfe'. At Camberton in Cambridge 'a spire Steeple in the Skye' was seen, and at Brandon in Norfolk a 'Pike or Lance' was seen to descend from 'Heaven'.

*an octagonal spire.

The 'ball of wilde-fire' which 'fell upon the earth' then 'burnt up and spoyled about an Aker of Graine' at Sopham in Cambridge, was possibly ball-lightning associated with UFO activities.

The 'Navie or Fleet of Ships in the ayre, swiftly passing under Sayle, with Flags and streamers hanged out, as if they were ready to give encounter' seen at Brandon, is self-explanatory regarding UFO reports, but the 'Flags and streamers hanged out . . .' might have been a corona or coloured effects from electrical discharge trailing behind the UFOs like a jet contrail, or a comet, meteor or meteorite tail. The 'pillar' or 'Cloud' seen 'to ascend from the earth, with the bright hilts of a sword towards the bottom of it . . .' i.e. 'Betwixt Newmarket and Thetford' might also have been an electrical effect or corona?

In the Counties of Norfolk, Suffolk and other Eastern parts of Britain in that area, 'Thunder' 'a whole Regiment of Drums' and 'the like Military sound . . .' was heard that day. Was this sound UFOs going through the sound barrier? Or was it rocket motors for terrestrial atmosphere, starting and stopping? ('with perfect notes and stops')?

'In all these places there was great Thunder, with Raine and Hailstones of extraordinary bigness and round, and some hollow like rings.' Were electrical activities of UFOs disturbing the atmosphere, or were clouds being artificially 'seeded' with chemicals to create rain and hail for some purpose?

It would be difficult to form a definite opinion of what was *really* happening on that 1646 afternoon. Analysis of the 1646 report indicates that UFOs and pilots *were* conducting scientific activities on that day. What these were, could be an interesting subject fcr discussion. Perhaps Earth's technology is not yet sufficiently advanced for scientists to understand the full significance of this mystery? Excavations around the sites mentioned in the 1646 report, might disclose scientific devices unknown to us in use and purpose?

Standing stones emitting ultra-violet radiation and up to a 32 km 180 m distance at a 23½° angle, or multiples of 23½°, add to the mysteries of ancient Britain. What was, or is, the purpose of all these obscure recondite facts? Were UFOs using a section of this 'power network' to recharge in 1646?

It is engaging to note that nearly 118 years after the 1646 Signes from Heaven report, during March 1764, 'Speares of Light' were seen in N.E. England!

13. ST. MICHAEL SPOKE TO JOAN OF ARC FROM A 'BRILLIANT LIGHT' IN THE AIR

'There are more things in heaven and earth, Horatio,/Than are dreamt of in your philosophy.'

(*Hamlet I.v. 166.*)

Did Joan of Arc, born 1412, converse with space beings? According to one curious legend, Joan's birth was prophesied by Merlin in the fifth century.

At the age of twelve, in 1424, Joan first experienced the strange event of being spoken to by St. Michael the Archangel, from a 'brilliant light' in the air. Several times after, she had the same experience; i.e. the Archangel always spoke to her from the 'great light' in the air.

One day, St. Catherine and St. Margaret appeared to her; both had on their heads 'crowns of gold'.

After four years of conversing with the saints, Joan was commanded by St. Michael to help her country.

Was the 'brilliant light' a UFO, the 'gold crowns' space helmets, and St. Michael, St. Catherine and St. Margaret space beings who guided Joan in her mission?

New historical research seems to prove that Joan was not burnt at the stake by the English at Rouen in 1431; this may have been a legend instigated for political reasons. The French Armoises family have for centuries possessed documents which state that Joan was not burnt, but imprisoned for several years. When released, she moved close to Metz, the capital of Moselle; there she married Robert des Armoises. Further research by genealogists almost conclusively proves that Joan was of royal blood, being the daughter of Queen Isabeau and the King's brother Louis d'Orleans. She probably did not have the humble peasant birth as recorded in history books.

Compte Pierre de Sermoise, the last descendant of the Armoises family, has amassed a large volume of evidence in connection with these new discoveries.

14. AN ITALIAN RENAISSANCE POET DESCRIBED UFOs AND PILOTS

During a 5th October, 1968 Press conference at Dallas, Col. James A. McDivitt, Command astronaut on the Apollo 9 Lunar Test Mission, said that one of three UFOs was photographed in space during the June 1965 orbital flight, and that NASA cannot identify it. Col. McDivitt said:

'They're there, without a doubt, but what they are is anybody's guess.'

Ludovico Ariosto (1474–1533), the most famous poet of the Italian Renaissance, wrote of flying ships, chariots, and cars, piloted by 'demons' or 'wizards', in his greatest work, the renowned *Orlando Furioso*, first published in 1516.

Was he describing UFOs and pilots? Just a few years earlier, another eminent Italian Piero della Francesca (1416–92) painted an extraordinary fresco in the church of San Francesco D'Arezzo in Tuscany, Central Italy. This fresco portrays what are undoubtedly UFOs in flight, complete with cabins on top of the obliquely-viewed (lenticular-shaped) bases. The UFOs in this painting closely resemble those photographed and reproduced in UFO books!

The following excerpts from Ludovico's *Orlando Furioso*, if written in modern terminology, would most definitely indicate UFO sightings!

In Canto I, Stanza VIII he speaks of:

> 'proud demons sailing the heavens in great ships of glass'.

Did he really mean spacemen piloting their UFOs through the skies?

Here is a similar description:

Canto II, Stanza LII

'Up to the starry sphere with swift ascent the wizard soars.'

In the following stanza Ludovico describes how an aerial 'demon' or 'wizard' can manoeuvre his flying 'horse' so skilfully.

Canto III, Stanza LXVII

'His swift horse is taught to wheel,
And caracol and gallop in mid sky.'

Was he describing the manoeuvres of a UFO seen in the Renaissance?

Canto IV, Stanza IV states that:

'With eyes upturned and gazing at the sky,
As if to witness a comet or eclipse.
And there the lady views, with wondering eye,
What she had scarce believed from others' lips
A feathered courser, sailing through the rack,
Who bore an armed knight upon his back.'

Canto IV, Stanza V

'Broad were his pinions, and of various hue;
Seated between a knight the saddle pressed,
Clad in steel arms, which wide their radiance threw,
His wonderous course directed to the west:
There drop't among the mountains lost to view.'

Canto IV, Stanza VI

'He sometimes towering, soars into the skies;
Then seems, descending but to skim the ground . . .'

In the last three stanzas, Ludovico describes the sighting of a 'feathered courser' which 'sailed through the rack' or wispy clouds moving through the sky. The 'feathered courser' had wings or feathers of 'various' colours, and was driven by an 'armed knight'. Ludovico then describes the manoeuvres of the 'feathered courser'; i.e. ascending, descending, and skimming the ground.

Stanza IV states that the lady saw one of the 'feathered coursers' for the first time, after scarcely believing the reports of others who had seen them; this very same parallel is to be found in modern UFO reports!

Stanza V relates that the 'knight' was 'Clad in steel arms, which wide their radiance threw.' Was Ludovico referring to the shimmering radiance of a UFO pilot's metallic space suit? And the 'feathered courser', was it a UFO?

Canto IV, Stanza L describes the flight of a 'knight', 'demon' or 'wizard' in an aerial vehicle:

> 'His course for where the sun, with sinking light,
> When he goes round the heavenly crab, descends;
> And shoots through air, like well-greased bark and light.'

The 'heavenly crab' is the constellation of Cancer. 'Bark' means either barque or bark, a sailing ship. Many scribes of old recorded sightings of aerial objects as ships sailing through the air.

Canto XXXIII, Stanza LXXXIV makes mention of a strange 'bird' 'three yards' long, with 'black feathers' and a 'fiery eye'. 'And like a ship sails, two spreading pinions shook.'

Canto XXXIII, Stanza LXXXV

> 'Perhaps it was a bird; but when or where
> Another bird resembling this was seen
> I know not, I, nor have I any where,
> Except in Turpin, heard that such has been,
> Hence that it was a fiend, to upper air. . . .'

What was the strange 'bird' with 'fiery eye' and wings like 'ship sails', which was 'a fiend to upper air'? A UFO?

Canto XXXIII, Stanza CXII mentions a flying steed: 'Of winged horse arriving through the air.' A UFO?

Canto XXXIII, Stanza CXIV says:

'And there with spacious wheels, on earth descended:
The king, conducted by his courtly crew.'

Did he mean that UFOs landed and spacemen alighted?

The following three stanzas describe the journey of a 'knight' to the vicinity of the moon in a flying 'chariot'.

Canto XXXIV, Stanza LXVIII

'A chariot is prepared, erewhile in use
To scour the heavens . . .'

Canto XXXIV, Stanza LXIX

'Who when the knight and he well seated are,
Collect the reins; and heavenward they aspire
In airy circles swiftly rose the car.'

Canto XXXIV, Stanza LXXX

'The chariot, towering, threads the fiery sphere,
And rises thence into the lunar reign.
This in its larger part they find as clear
As polished steel, when undefiled by stain;
And such it seems, or little less, when near,
As what the limits of our Earth contain:
Such as our earth, the last of globes below,
Including seas, which round about it flow.'

Did a Renaissance knight journey to the Moon hundreds of years before US astronauts? Or was the 'knight' a spaceman, and the 'chariot' a UFO?

Did Ludovico Ariosto describe UFO sightings in his poetry?

The Italian Renaissance produced the most august assemblage of creative geniuses in any period of Western history; such a tremendous outflow of creative vigour and intensity was not matched before or since. Perhaps the intellectual stimulus of the Renaissance was ideal for the rebirth of so many eminent souls?

15. JONATHAN SWIFT DESCRIBED ELECTRONIC INSTRUMENTS IN A UFO CABIN

'The most stimulating theory for us is that UFOs are material objects, which are either "manned" or remote-controlled by beings who are alien to this planet . . .'

'It is suggested that "at least three, and maybe four different groups of aliens" (could be visiting Earth).

'We should not deny the possibility of alien control of UFOs on the basis of preconceived notions.'

> From Volume II of *Introductory Space Science*, the US Air Force Academy (Colorado) textbook. Publicized in syndicated Press article, January 1971.)

UFO authors have speculated at how Jonathan Swift was able to describe so accurately the two tiny moons of Mars – Phobos and Deimos – 151 years before their discovery by the Naval Observatory in Washington. The mystery is solved by a detailed analysis of the story 'Voyage To Laputa' published in *Gulliver's Travels*.

In 'Voyage to Laputa' Swift gives an accurate account of a flying saucer – or 'flying island' – and the inhabitants, who told Gulliver about the Martian moons in, according to the story, early February 1708.

It seems obvious that *someone* (who, and when?) had been a guest on board a UFO!

The full story from 'Voyage to Laputa' is too long to quote here but the following is a condensed analysis.

(1) In Chapter 1, Gulliver described the 'flying island' as being 'a vast opaque body', 'smooth and shining very bright', 'inhabited by men, who were able (as it should seem) to raise or sink, or put it into a progressive motion, as they pleased'. The 'island' also 'seemed for a while to stand still'.

Every one of the factors have been recorded in modern UFO sightings!

Gulliver mentioned that 'In the lowest gallery I beheld some people fishing with long angling rods, and others looking on.'

Were the 'long angling rods' antennas?

Gulliver further noticed that some of the 'flying island' people 'seemed to be persons of distinction, as I supposed by their habit' ('habit' or space suit?).

(2) Swift said that Gulliver 'could plainly discover numbers of people moving up and down the sides of it'. After waving, calling and shouting, 'with the utmost strength of my voice' the men finally noticed Gulliver, and 'at length one of them called out in clear, polite, smooth dialect, not unlike in sound to the Italian'.

This phenomenon was closely paralleled on 26th June, 1959, between 6.45 p.m. and 11.04 p.m. at Boianai, Papua, when the Rev. Father W. B. Gill of the Anglican mission, along with 37 other witnesses, waved to four beings standing on the deck of a UFO hovering about 90 m up in the air.

(3) In Chapter II, Gulliver on the 'flying island' said: '... At last we entered the palace, and proceeded into the chamber of presence, where I saw the king seated on his throne, attended on each side by persons of prime quality. Before the throne was a large table filled with globes and spheres and mathematical instruments of all kinds. ...'

Was the 'chamber of presence' the control room of a UFO and 'the king seated on his throne' the chief UFO pilot? What of the 'table filled with globes and spheres and mathematical instruments of all kinds ...'; were these UFO instruments?

(4) Gulliver said about his tutor, that:

> 'He gave me the names and descriptions of all the musical instruments, and the general terms of art in playing on each of them.'

He then states:

> '. . . On the second morning about eleven o'clock, the king himself, in person, attended by his nobility, courtiers, and officers, having prepared all their musical instruments, played on them for three hours without intermission, so that I was quite stunned with the noise; neither could I possibly guess the meaning, till my tutor informed me. He said the people of their island had their ears adapted to hear the music of the spheres, which always played at certain periods. . . .'

It seems most doubtful that the 'flying island' people were playing musical instruments for 'three hours without intermission', but, they '*had their ears adapted to hear the music of the spheres, which always played at certain periods. . . .*'

(my italics)

What the 'flying island' people may really have been doing, was receiving on electronic instruments what is known as 'the dawn chorus'! The US National Academy of Sciences reported in 1962 that strange music from space had been detected, and was received shortly before dawn, or just after. The report stated that these electronic sounds were: 'a series of short, distinct musical tones, either rising or falling in pitch, and often overlapping in time'. An alternative possibilty is that the 'flying island' people were in radio communication with other planets or spacecraft. Swift could not have known of electronics in the eighteenth century; or could he?

(5) Chapter III mentions that 'twenty lamps' 'continually burning' 'cast a strong light into every part'. Was this an artificial lighting source? If so, an eighteenth-century inhabitant of Earth would undoubtedly have great difficulty in understanding and describing such a weird and wonderful sight.

(6) A detailed description of the 'loadstone' 'upon which the fate of the island depends' is given. Loadstone, or magnetite, is a magnetic mineral comprised of iron oxide.

A drawing built up from data in Chapter III show, in cross-section, a strikingly similar construction to cross-section drawings of UFOs in certain books; e.g. *Space, Gravity, and the Flying Saucer* by Leonard G. Cramp.

Several UFO authors have described a central 'magnetic pole' in UFOs which reaches from floor to ceiling; this is supposedly an essential and integral component of the UFO. But Swift forwards a very detailed account of the magnetic pole; the 'loadstone of a prodigious size, in shape resembling a weaver's shuttle'. *How could he have known?* The narration then relates that the 'loadstone' or 'Magnet' has an axle of 'adamant passing through its middle, upon which it plays, and is posed so exactly, that the weakest hand can turn it'. He said that the loadstone 'is hooped around with a hollow cylinder of adamant . . . and supported by eight adamantine feet . . .'

Swift was attempting to describe a translucent, transparent or coloured, gem-like substance. Adamantine Spar (Corundum) is a mineral comprised of aluminium oxide. It is either colourless, or found in colours of blue, green, red, topaz and violet.

After describing the function of the 'loadstone', Swift describes how the 'island' rises, descends, and moves from one place to the other. He illustrates this point with a map, and a drawing of the 'flying island'. Swift also mentioned that Gulliver was 'not in the least sensible of the progressive motion made in the air by the island . . .' If it was a UFO powered by magnetic forces, he probably would not be aware of any movement!

(7) Gulliver observed that the inhabitants of the giant 'flying island' had 'glasses far exceeding ours in goodness. For, although their largest telescopes do not exceed three feet, they magnify much more than those of a hundred yards among us, and at the same time, shew the stars with greater clearness.'

This is possibly achieved with electronic image multiplication; in 1963, an image-intensifying device that amplifies starlight 100,000 times was used for the first time at Kitt Peak National Observatory, Arizona.

(8) Swift said that the 'flying island' people told Gulliver that the inner Martian moon (Phobos) revolved around Mars in 10 hours; today it takes 7 hours 39 minutes. This validates certain scientific discoveries that Phobos may be gradually moving closer to Mars. Swift's figures were accurate over two hundred years ago, when Earth's astronomers could not see the tiny moons!

Calculations from information in *Gulliver's Travels* indicate that the outer moon Deimos has actually moved about 6,436 km further out from the centre of Mars since described by Swift. This behaviour may possibly indicate that both moons were not ejected from Mars during the formative period.

16. A SPACEMAN ALIGHTED FROM A GLOBE-SHAPED UFO IN 1790

'Intelligent beings abound in the universe, and most of them are far older than we are.'

(Dr. W. Howard, Harvard University.)

Alberto Fenoglio of Turin discovered the following fascinating report during UFO research, and published it in *Clypeus, Anno III, No. 3*. The report is from Liabeuf, a Police Inspector, who in June 1790 was despatched from Paris to investigate a mystery close to Alençon:

'At 5 a.m. on 12th June, some peasants observed an enormous globe which seemed to be surrounded by flames. At first they thought it might be a Montgolfier balloon on fire, but its great speed, and a whistling sound coming from it, puzzled them. The globe slowed down, made a rocking motion, and then dashed on to the top of a hill, uprooting the vegetation growing on the slope. The heat which emanated from the object was so great that the grass and shrubs caught alight afterwards.

'The peasants managed to isolate the fire which might otherwise have spread over the whole area. By evening the globe was still warm, and there occurred an extraordinary – indeed not to say unbelievable – thing.

'The eyewitnesses of this event were two mayors, a physician, and three other local authorities who confirm my report, not to mention the dozens of peasants who were present.

'The sphere which was large enough to have contained a carriage, was intact after all this flying about. It had aroused such curiosity that people came running from all directions to see it.

'Then suddenly, a sort of door opened, and there came out a person, just like us, but dressed in a strange manner, in clothes adhering completely to the body, and, seeing this crowd of people, the person murmured something incomprehensible and ran into the wood.

'The peasants backed away instinctively, in fear, and this saved them, for shortly afterwards the sphere silently exploded, throwing pieces in all directions, which pieces were consumed until reduced to powder.

'Searches were undertaken to find the mysterious man, but he seemed to have dissolved in thin air, for up till now not the tiniest trace of him has been discovered. Unless he has vanished from our plane of existence so as to leave behind no trace of himself.

'Was this a being who had come from another world in this strange means of conveyance? I am no savant; but such is the idea that has suddenly come into my mind. ...'

This report was sent to the Paris Academy of Sciences in June 1790 but was treated with disbelief. Possibly a more serious interest would be accorded this incident if it was to happen in the twentieth century.

'... clothes adhering completely to the body. ...' Did the Police Inspector report a being in a space suit?

17. DID A GIANT SPACESHIP ABDUCT A BRITISH REGIMENT IN 1915?

'The UFOs really exist, and apparently come from other planets.'

(Quote from Javier Garzon, a physicist on the staff of the National Astronomical Observatory, Mexico City.)

An incident which happened during the Dardanelles Campaign in Turkey seems incredible, but has been substantiated by several eye-witnesses. What happened on 21st August, 1915? Close to Hill-60 or Kaiajik Aghala, inland south of Suvla Bay on the Gallipoli Peninsula, a British regiment, comprised of a few hundred men, marched into a strange pale greyish 'cloud' which appeared to eye-witnesses to be settled over Kaiajik Dere – a depressed area in the ground.

None of the men came out of the 'cloud' again, but after approximately one hour, the curious 'cloud' lifted off the ground, rose about 1·2 km until it met with 6–8 similar 'clouds', then all moved off in the direction of Bulgaria in Eastern Europe.

The 'clouds' were all elongated in shape; the 'cloud' on the ground was estimated to be 241 m in length, 60 m in height and width, and of a 'dense' 'solid' appearance.

Eye-witnesses to the incident are: 'ANZAC veterans 4/165 Sapper F. Reichart of Matata, Bay of Plenty, NZ, who recorded an account, 13/416 Sapper R. Newnes of Cambridge, NZ, and J. L. Newman of Otumoetai, Tauranga, NZ, who signed this same account.

Reichart remarked that after the surrender of Turkey in 1918 Britain demanded the return of the missing regiment:

'Turkey replied that she had neither captured this

regiment, nor made contact with it, and did not know that it existed.'

One other notable point is mentioned in the Suvla Bay landing by John Hargrave, who commented that on 21st August, 1915, several battalions lost direction in this area due to the compass needle 'inclining too much to the north'.

Were there magnetic-powered spaceships in the vicinity? What *really* happened?

18. DOES A MILKY WAY, TRANS-GALACTIC SPACESHIP BEACON NETWORK EXIST?

What is the newly-discovered 'W-3'? Far, far from Earth on the outskirts of our Milky Way galaxy are five mystery sources, forming the shape of the Greek letter (Ω), of unnaturally intense MASER* radio energy.

A weak radio noise on the 21 cm or 1,420·405 mega-cycle-per-second band is heard throughout the universe. This 'steady drone' or 'song of the universe' is a radio emission from neutral hydrogen clouds in space. This frequency, postulated in 1945, discovered in 1951, is one of the focal points of attention in listening for intelligently modulated radio signals from spacemen; it is considered as the logical frequency on which to superimpose messages.

In 1943 Dr. Kaj Aa. Strand, director of the US Naval Observatory, discovered at Sproul Observatory, Swarthmore College, a planet orbiting around the binary star 61-Cygni, magnitude 5, 11·1 light years from Earth in the constellation Cygnus. Until recently scientific belief was that planets in orbit around a binary star would have an unbalanced orbit and be unable to support inhabited planets, but Dr. Su-Shu Huang of NASA calculated that inhabited planets can exist around a binary if the planets are at sufficient distance from the centre of the primary and in a circular orbit.

A new Soviet hypothesis is that fluctuations in the intensity of spectral lines from Cygnus and other constellations, previously thought to be from natural causes, may, in reality, be light signals sent by intelligent beings.

Dr. H. C. Ko of the Ohio State University Radio Observatory in Columbus, reported in the early 1960s that unusual radio emanations had been received from Cygnus. These separate, and distinct, radio signals were

*microwave amplification by stimulated emission of radiation.

received on a 600 megacycle frequency. The new radio source is called Cygnus-Y and is located 8° north-east of a previously known radio source Cygnus-X. In this context it may be important to note that the following stars in Cygnus have planets in orbit: 61-Cygni A, Ci 2347, Ci 2354, and Ci 1299.

Early in 1955 radio astronomers at Princeton, NJ first detected bursts of radio energy on the Seneca radio telescopes. These emissions on 13·5 and 10 m frequencies, for many years a mystery, were discovered to be at least partly radiating from one of Jupiter's moons, by Dr. Keith Bigg, a scientist at CSIRO.

In September 1956 while Mars was at its closest approach to Earth, the US Navy received short-wave radio signals (from Mars) on a 3 cm wavelength. These signals received on the Naval Research Laboratory's 15·1-m-diameter radio telescope were assumed to be from natural causes. But in 1959, Professor Anotil, a distinguished Soviet radio astronomer from Irkutsk, announced to the *Science Press* of Moscow that systematic radio signals sent from Mars are *not* natural phenomena, but are signals broadcast by *intelligent beings* attempting to communicate with other planets!

In November 1957 scientists detected highly unusual radio signals on a 14·286 megacycle frequency, coming from a moving source in space! The US did not have any satellites in orbit at that time; the only two Russian satellites broadcast on frequencies of 20·005 and 40·002 megacycles. Did the mystery signals originate from an alien spaceship, or an unidentified extra-terrestrial satellite placed into orbit around Earth?

At the end of 1965, University of California radio astronomers detected radio signals from space which had patterns totally different from anything in the radio spectrum. No known natural interstellar gas clouds, or natural elements, could emit signals such as these, which fluctuated in time scales of hours, or less. The unidentified signals were called 'mysterium'.

During 1966 Professor D. Martinov, director of the Astronomical Observatory of Moscow University, announced that there are *probably* other civilizations of intelligent beings in space, and that the USSR was 'eager to co-operate with other countries' to attempt to receive signals from the beings. The Soviet Union found the task too difficult to undertake all work by itself.

Evidence of increasing interest among Soviet scientists to attempt to receive intelligently modulated radio signals from spacemen, is indicated by the following:

> 'One of the world's most famous observatories, Pulkovo, is installing special, extremely powerful instruments, to seek for contacts with civilizations in outer space.
>
> ' "This is one of the most interesting tasks facing astronomy," comments Academician Aleksandr Mikhailov, President of the Academy of Sciences Astronomical Council.'
>
> (*Soviet News*, 26th September, 1969.)

Dr. Robert Jastrow, director of the Goddard Institute for Space Sciences at NASA, believes that beings in other solar systems may have been listening to our radio broadcasts for many years.

Strange echoes of radio signals sent from Earth have been *returned* from space on several occasions. Commenting on this in 1964, Dr. M. A. Mercer, physics lecturer at Southampton University, said:

> 'There is the case of some long-delayed echoes in a radio transmission investigation about 30 years ago which has never really been explained. It is entertaining to speculate that they might have been picked up by an exploring probe* which has relayed the information back to its planet, many light years distant.'

*Now calculated by Scottish theorist Duncan Lunan to be an extra-terrestrial probe permanently stationed in the vicinity of the Moon; which arrived circa 11000 B.C., and was sent from a planet of the binary star Pulcherrima 103 light years away.

PULSARS MAY BE TWO-MILLION-YEAR-OLD SPACESHIP BEACONS

Curious extra-terrestrial radio signals were detected early in 1968 by scientists at Cambridge University. Radio astronomers at Arecibo, Puerto Rico, Parkes, in the USSR, and in other countries, all study the now famous 'Pulsars' which emit the unusual signals.

Cambridge scientists discovered four sources of Pulsar signals, calculated to be coming from Pulsars approximately 200 light years from Earth, and within our galaxy. Some Pulsars may even lie beyond the periphery of the galactic disc, which measures 100,000 light years in length, with a diameter of 30,000 light years (or 30M parsecs long, and 10M parsecs across).

The strange 'lighthouse' signals arrive in pulses, one exactly every 1·3372795 secs for a period of one minute, then they vanish for 3–4 minutes, reappear for one minute, and continue the same cycle all over again.

The Pulsar signals repeat so precisely, one every 1·3372795 secs, that they have an accuracy of one point in 100,000,000. This is *far* greater precision and consistency than could be achieved with any Earth-made electronic equipment! The accuracy *exceeds* that of the US Naval Observatory radio station WWV which broadcasts time signals, but equals that of atomic clocks!

On 15th June, 1968, the fifth Pulsar was discovered, by US radio-astronomers operating the 90½-m-diameter radio-telescope at the National Radio Astronomy Observatory (NRAO) in Greenbank, West Virginia. This Pulsar, designated HP 1506, emits one pulse every 0·7397 secs, and has an accuracy to within 1/10,000 of a second. Another – PSR 0833 – pulses extremely rapidly every ·089 secs, the fastest, at ·033 secs. Many new Pulsars were discovered, and by mid 1973 100 had been catalogued.

Undoubtedly the most dramatically unusual Pulsar was discovered in the Milky Way's northern constellation Hercules, in November 1971. This enigmatic

X-ray source pulses every 1·24 secs for nine days, abruptly shuts off for 27 days, then recommences the same cycle, indefinitely.

Dr. Frank Drake of Cornell University said about the Hercules Pulsar:

> 'This discovery is every bit as bewildering as the finding of the first Pulsar five years ago. There is no known reason why a star should disappear and then reappear at such regular and predictable times.'

Astronomers endeavoured to discover natural (causes) for the unusual behaviour of Pulsars, but as soon as theories were published, the Pulsars exhibited new or different characteristics.

The *New York Post* said:

> 'Prof. Alan Barret of the Massachusetts Institute of Technology has wondered out loud if these signals could be part of a "vast interstellar communications network" on which we have stumbled.'

Perhaps another possibility could be that the Pulsars are linked in a vast power interstellar transmission network?

The most intriguing thoughts on the subject are that the Pulsars are powerful navigation beacons for interstar spaceships, for the obvious purpose that spacecraft crews don't become lost in the vastness of space!

Even if, as theorized, the Pulsars are pulsating – rapidly rotating – neutron stars; a 'third class of star' (etc.), it still will not alter the possibility that they may have been artificially developed for the (probable) purpose of navigation beacons!

Several highly esteemed Soviet scientists believe there may be superior, higher, outer-space communities who have progressed to a technological level so far beyond our comprehension, that they are actually able to influence developing stars to the point where they can guide, and/or change their development! Another example of this thought is extra-terrestrials exploding stars into Super Novae with LASER, etc., beams! And are Pulsars – neutron stars – the collapsed, condensed, massive core remnants of stars deliberately imploded by extra-terrestrials, to create navigation beacons, or whatever?

A relevant observation for the possibility of artificially-developed Pulsars is that apparently they are about the same age; i.e. about 2,000,000 years old. It does not seem possible that so many star objects would be about the same age, especially as they are *not* all clustered together, but *spread throughout our galaxy*! Although a curious concentration of Pulsars is situated along the *plane of the galactic equator*!

19. GEORGE ADAMSKI SAW OTHER WORLD LASER HOLOGRAMS NINE YEARS BEFORE EARTH SCIENTISTS INVENTED THIS TECHNIQUE

'In the distant future we will encounter some other intelligent life in our solar system.'

> (From a speech by former astronaut Frank Borman, at Pierre, South Dakota, 28th January, 1971.)

Whatever has been thought, written or talked about George Adamski's UFO books, one notable point in his works cannot be denied. Throughout his books he described various phenomena, etc., in space, which he supposedly personally witnessed on, and from, extra-terrestrial spaceships; i.e., the 'fireflies', etc., which were not seen by Earthmen astronauts and cosmonauts until many years later!

With the development of LASER Holograms in the 1960s, further support is given Adamski's earlier assertions; for example, in *Inside The Spaceships* Adamski remarks on page 212, in reference to being shown Venusian scenes on board the spaceship in 1954, that:

> 'I was delighted at the prospect of such a travelogue, and wondered on which screen it would appear. But there was no screen. Before my astonished gaze, as the lights dimmed, the first scene hung suspended upon the space of this room!
>
> 'Orthon seemed to enjoy my amazement, and explained:
>
> ' "We have a certain type of projector that can send out and stop beams at any distance required. The

stopping point serves as an invisible screen where the pictures are concentrated with colour and dimensional qualities intact."

'The scene at which I was looking, seemed, in fact, so definitely "there" that it was with the greatest difficulty I could believe myself still on the ship . . .'

What George Adamski was describing, was a highly developed and sophisticated LASER Hologram projection system. With the development of the first LASER in 1960, the first LASER Hologram was shortly to follow (in 1963), but Adamski described an advanced LASER Hologram projection system *nine years* before the first LASER Hologram was invented on Earth!

As Adamski described, in Holography images are projected on LASER beams which can be halted and (by diffraction grating) focused in mid-air, the atmosphere itself serving as the screen. The Hologram image is three-dimensional, and hangs suspended in mid-air without the necessity of a reflective screen. Colour Holograms up to 60 cm high, and movie sequences, have been developed.

The image is first photographed and recorded on emulsion in patterns, etc., with slow argon, helium, or high-speed ruby Q-switched LASERS. The image can also be recorded in alkali-halide potassium bromide crystals.

If Adamski did in fact see a LASER Hologram projection system on board an alien spacecraft, it could be rather pointed that Earth scientists managed to develop an identical Hologram system (still under development) by utilizing LASERS in a way which *so closely* matches Adamski's description some years earlier!

Of possible significance in this connection, is a short passage in Adamski's *Flying Saucers Farewell* in which he says, on page 147, that while on a train to Weston-super-Mare where he was to lecture on UFOs and space people, he met an unusual man . . .

'After a short interval, the man in my compartment started a conversation. To my amazement, he was a spaceman working as a scientist for the British government! He, and countless others like him, are working on various scientific projects for every government in the world.'

Adamski's statement is paralleled by Dr. Hermann Oberth, the world-famous aerospace and rocket authority!

20. SCIENTIFIC PROOF OF EXTRA-TERRESTRIAL LIFE

Dr. Hermann Oberth, internationally-known rocket pioneer and space authority, UFO lecturer, and head of the US CALTECH Laboratories until 1955, said:

'We cannot take all the credit for our record advancements in certain scientific fields alone; we have been helped!'

When asked by whom, he replied:

'The people of other worlds!'

This chapter, devoted mainly to scientific discoveries relating to extra-terrestrial life, plus other disclosures of a scientific and historical nature, further prove that space beings have been visiting Earth since antiquity.

Dr. Rainer Berger, a chemist at the Convair Scientific Research Laboratory in San Diego, simulated conditions which apparently exist in space. By freezing a methane, ammonia, and water admixture to a specified sub-zero temperature, then exposing the frozen synthesis to a proton stream for 200 seconds in a cyclotron, he obtained acetamide acetone, and urea, the basic building blocks of organic matter.

A proton is a positively charged unit in the nucleus of every atom; in space cosmic radiation is comprised of 85 per cent protons. Dr. Berger's experiment indicated that organic matter is being produced in space under similar conditions.

More than 23 outer-space molecules associated with life-forms have been detected since 1968; the most recent are thioformaldehyde, acetformaldehyde, and formaldimine, detected by scientists at Parkes.

After probing the heavens for years to locate chemical molecules associated with life-forms, US radio-astronomers made some interesting discoveries. The first was of a deep-space combination of hydrogen and oxygen atoms – hydroxyl. In December 1968 University of California radio-astronomers announced the discovery of ammonia in the interstellar hydrogen clouds, and in February 1969, the discovery of water.

In March 1969, scientists operating the 42-m-diameter radio-telescope at NRAO announced the discovery of formaldehyde. Formaldehyde, a colourless acrid gas, or an organic substance found in the chlorophyll cells of plants, is 2 hydrogen atoms and 1 oxygen atom bonded to carbon, or H.CHO. Only methane has not been discovered due to its molecular structure, which cannot be detected by radio-telescopes. But methane almost certainly exists in outer space and it is chemically related to formaldehyde.

In 1952, graduate student Stanley Miller at the University of Chicago, succeeded in producing amino acids – essential components of protein structure – in an atmosphere of methane, ammonia, hydrogen and water bombarded with electricity for seven days.

Dr. David Buhl of NRAO said that nine significant formaldehyde areas in space were probably the early processes of formation for nine future civilizations. (Or future homes for reincarnating souls when their own solar systems had decayed or come to an end?)

In August 1970 Professor Zdenek Kopal of Manchester University announced that astronomers were watching for the first time, new planets forming around the giant star Epsilon Aurigae, 20 times the size of our sun. Will this new planetary system 5,000 light years from Earth, eventually support intelligent beings?

Also in 1970, the Rome meetings of the International Astronomical Union told astronomers that recent discoveries of discreet sources of X-ray protons may date back to early in the birth of the physical universe. (But we still don't know how the first material hydrogen atoms or atom came into existence from nothing!)

The following three stars less than 12 light years from Earth, are believed to have possible solar systems supporting intelligent beings.

EPSILON ERIDANI	10.8 light years from Earth
EPSILON INDI	11.3 light years from Earth
TAU CETI	11.8 light years from Earth

The next three interesting stars within 13·5 light years from Earth have actually been discovered to have planets in orbit – (are any inhabited?).

The star LUYTEN 726–8 (magnitude 12·5), 7·9 light years from Earth in the Constellation CETUS was discovered by Dr. Willem J. Luyten in 1948. In 1969, Dr. L. W. Fredrick and graduate student P. J. Shelus working at the Leander McCormick Observatory at the University of Virginia discovered a planet in orbit around this star.

In 1960, Dr. Sarah Lee Lippincott at Sproul Observatory discovered a planet in orbit around the star LALANDE-21185 (magnitude 7·5), 8·2 light years from Earth in the Constellation CANES VENATICI.

Of unusual interest is the star ROSS-614 (magnitude 10·9), 13·1 light years from Earth in the Constellation ORION, aptly named 'The Hunter'! In 1955 a planet was discovered in orbit around this star. Legends from antiquity speak of 'evil' space beings in ORION solar systems who once attempted to destroy the PLEIADES!

A USSR theory is that the 'Cepheid Variables' or 'winking stars', may, in reality, be gigantic heliographs signalling in code to communities of space beings!

The Arabs who named four times as many stars as other ancient astronomers, had an aura of mystery surrounding some of their discoveries. The tenth-century Arab astronomer Abu 'l Wafa described the deviation of the Moon from its orbit, due to solar influence at different points; he computed these variations precisely, but *did not* possess any delicate, precision instruments to enable him to do so! Did he learn these astronomical discoveries from spacemen?

Sir Bernard Lovell, director of the Jodrell Bank Radio Observatory, said while on a visit to the Soviet Union in 1963, that there are:

'many communities of other beings in different parts of the universe'.

During a 1966 public lecture at Jodrell Bank, Sir Bernard Lovell said:

> 'present indications showed a very high degree of probability that life forms might have emerged elsewhere in the universe.
>
> 'Man must give up the idea that he is unique . . .'

Another possible indication of Earthman being taught science by space beings in antiquity, may be in this news item from the *New York Times*, 8th January, 1950, referring to the discovery of a clay 'textbook' in Iraq:

> 'Schoolboys of the little Sumerian county seat of Shadippur about 2000 B.C. had a "textbook" with the solution of Euclid's classic triangle problem, seventeen centuries before Euclid . . .
>
> 'It suggests that mathematics reached a stage of development about 2000 B.C. that archaeologists and historians of science had never imagined possible.'

DID A UFO PILOT JAM THE TIDBINBILLA SIGNALS?

Can we *really* be certain of *all* scientific data transmitted back to Earth from satellites?

The two 1965 meteorological satellites Tiros and Nimbus took thousands of photographs of Earth from distances as close as 418 km 340 m to 933 km. The two satellites were expected to show 'objects' as small as 181 m in diameter, but out of the thousands of photos of Earth only three slight indications of civilization appeared. They were: Interstate Freeway-40 out of Memphis, Tennessee; what was possibly a jet contrail over the North Pole; and logging tracks in a Canadian forest. Only three slight indications of civilization . . . yet we *know* that Earth is teeming with life, and has enormous artificial constructions!

For many years there has, of course, been considerable controversy over whether the Martian 'canals' – 'canali' – do actually exist, but Dr. Clyde Tombaugh, the world-famous astronomer who discovered Pluto in 1930, and who at 10.45 p.m. on 20th August, 1949 saw a 'cigar-shaped' UFO at Las Cruces, New Mexico, said that he detected 'traces' of the Martian 'canals' in seven photos taken by Mariner IV in 1965. Mariner VI photographed a 'wide streak' near the equatorial region of Mars, while Mariner VII photographed a 'dark streak' 1200 km long by 160·9 km wide, midway between the Equator and the South Pole.

Mr. Eric Burgess, charter member of the British Interplanetary Society, leading authority on space exploration, and author of nine books dealing with space science, said about finding (intelligent?) life on Mars:

> 'The only way we can really find out is to get a man down there. Because with all the experimental probe vehicles they have to be programmed to do certain tasks; they're not flexible. Man is almost infinitely flexible.'
>
> (*The Plain Truth*. September, 1969)

Mr. Burgess predicted in 1939 that Man would land on the Moon before 1970.

Two enigmas associated with the Mariner IV Mars photographs: one was during July 1965 while picture signals of Mars were being sent back from Mariner IV; scientists at the Tidbinbilla tracking station were mystified as to why the picture signals were unexpectedly *jammed*! Approximately 40 km away a UFO was hovering due north of Canberra at a 1·5-km altitude. The UFO remained stationary from 10.50 a.m. to 11.30 a.m.

A NZPA press report, 19th July, 1965, said that the UFO

> 'glowed with a bright steady light, and did not appear to move until it suddenly disappeared – just before an RAAF aircraft sent to investigate reached the scene.'

When the UFO disappeared, the picture signals again started coming in at the tracking station!

The second mystery, or part of the first mystery, is that two Spanish researchers, Antonio Ribera and José M. Oliver, made the astounding discovery that photo No. 11

of a Mars crater broadcast back to Earth by Mariner IV *seemed identical* to the Clavius crater region of the Moon!

The two researchers drew maps of the Mars and Clavius craters and superimposed them; thirty-six similarities existed between the two craters! It is not possible for nature to duplicate the same crater formation in different parts of the universe!

Don Feliu Comella, Spanish mathematician, calculated that the odds *against* finding identical craters in different parts of the universe is 2128 followed by 53 zeros to one, or infinity!

The fact remains that these two craters seem identical, and have a similarity with each other exceeding 50 per cent. What is the explanation? Why were the signals jammed at the Tidbinbilla tracking station while the UFO was hovering nearby in July 1965? Did the UFO pilot electronically substitute a possibly slightly altered photo of the Clavius crater region of the Moon in place of a topographical feature or features, of Mars? Do spacemen not wish us to see something?

Another curiosity is the 'mystery haze' over the Martian South Pole which absorbs ultra-violet light radiation. UV photos show Mars as a 'white ball'. It is known that excessive ultra-violet light radiation will harm all life-forms; Earth is protected from dangerous amounts of UV radiation by the 'ozone layer'. Does the Martian haze contain, other than carbon dioxide, an unknown that excessive ultra-violet light radiation will harm thesis of gases, which protect possible life-forms from lethal or damaging quantities of the sun's ultra-violet rays? It is interesting to note that on 8th June, 1967, Princeton University astronomers detected a previously unknown – and at that time unidentified – gas, in the ultra-violet spectra of Venus and Jupiter!

Martian atmospheric elements detected by Mariner space probes, are oxygen, carbon dioxide, hydrogen, water vapour and water ice. And (traces, or quantity?) of ozone, a three-atom form of oxygen?

Dr. Sanford Siegel, US biologist, head of the

exobiology department at the Union Carbide Corporation Research Institute, New York State, announced that his research group discovered that several species of animals and plants developed increased resistance to freezing temperatures as the oxygen level was reduced. These experiments possibly simulate certain areas of Mars with (evidently) small quantities of (atomic – and/or molecular?) oxygen and low temperatures.

LIFE ON JUPITER, SATURN AND NEPTUNE?

Jupiter has been theorized to be possibly supporting life according to two top scientists at NASA. The scientists duplicated Jupiter's atmosphere, believed to consist of ammonia, helium, hydrogen, methane and neon, then bombarded it with artificial lightning to simulate solar radiation.

Dr. Cyril Ponnamperuma, head of the Chemical Evolution Branch at NASA, said that nine amino acids were produced in the experiment. Amino acids are the building blocks of protein, which in turn is the main element for construction of the cell's nucleus – deoxyribonucleic acid.

This experiment does not of course prove that life-form bodies did actually begin in this way, or in a similar way, on this or any other planet, solely by electro-chemical, proton-chemical, etc., chance reactions without the life-force intelligence. The above-mentioned experiment still lacks 14 amino acids required to form a complete protein molecule; but this and other experiments and discoveries of a similar nature do indicate likely planetary and outer-space chemical pre-conditions (planned by whom?) for eventual spiritual habitation of life-form bodily states.

Referring again to Jupiter, in 1960 the spin-rate increased, which was unusual. Jupiter normally has an unusually fast spin-rate for a planet of such mass, but according to 200 astronomers at a September 1965 meeting at Caltech, the increase in spin-rate *defied some basic physical laws.*

By 1965 the spin-rate had again reduced to its normal speed. What caused the spin-rate to increase, then *revert* to its normal speed?

Could any of Jupiter's 12 moons, or other planetary moons in our star system, support intelligent beings; if not on the surface, perhaps underground with artificial atmospheres?

Jupiter's two largest moons – Ganymede and Callisto – are half the size of Earth; the other two largest are Europa, and Io, which is the size of our own moon. These four Jovian moons are the most interesting, in addition to being the largest; their composition is of rocky and metallic minerals, while they have unusual 'surface markings', and maps have been drawn of these. Not a great deal is known of the other eight moons.

A radio astronomy probe from Parkes indicated that Saturn's surface is not frozen as thought, but moderately warm!

Another development was reported by Professor Audoin Dollfus, who while working at the Meudon Observatory in France, discovered a new moon of Saturn – Janus. On 15th December, 1966, three photographic plates recorded the very faint magnitude-14 satellite; this tenth moon is the closest to Saturn, but Janus may be a tenth Saturnian moon first discovered in 1904 by an astronomer at Harvard who named it Themis. Themis was then over nine times the distance of Janus from Saturn and was photographed several times before vanishing. Sixty-two years later a tenth moon was discovered close to Saturn. If Janus is in reality Themis, how could it shift position so drastically by natural laws?

A peculiarity associated with Saturn is its outer moon Phoebe which revolves in the opposite direction around Saturn from the other moons. This same characteristic applies to Neptune, outer Jovian moons, and the Martian moons. Are (any) of the moons just mentioned possibly artificial?

An interesting observation in this connection is associated with Neptune which possesses two satellites – 'Triton' larger than our moon, and innermost, revolves around Neptune in a retrograde or 'clockwise' motion – while the outer satellite revolves in direct motion.

Science News, 15th October, 1966, reported that Triton will crash into Neptune, or will be broken up, in 10,000,000 years, as recent calculations proved that *Triton is moving closer to Neptune*!

This is the same characteristic as artificial Earth satellites, many of which will be eventually attracted back to Earth by the gravitational pull and destroyed upon entering the atmosphere. Is Triton an artificial satellite?

HAVE SPACEMEN REPAIRED, AND CAPTURED, OUR SATELLITES?

Evidence indicates that on several occasions, spacemen may have repaired our malfunctioning satellites, and on one occasion, even captured a Russian satellite and did not return it to its orbit!

Look at the following facts:

(A) Blinking lights on the ANNA 'fire-fly' geodetic sphere satellite were observed in 1962 to fade until they eventually stopped. The blinking lights were completely out of action for many months until August 1963, when they very mysteriously resumed flashing!

(B) TELSTAR 1 and 2 both ceased transmitting, then after a period of inactivity long enough to indicate a permanent breakdown, they both (at different dates) resumed transmitting!

(C) Power on the Venus-bound MARINER II spacecraft cut off after apparently being struck by a meteorite – then suddenly returned!

(D) An 8th February, 1962, AAP press report said that:

> 'A mysterious electrical impulse prevented the RANGER-3 space-craft from taking television pictures of the Moon ... The mysterious impulse mechanically countermanded an "order" radioed about 200,000 miles from the Goldstone tracking station on January 28 ...'

(E) A Soviet communications satellite launched in October 1965 with an estimated life in terms of decades, mysteriously disappeared. Remarking on this strange anomaly, a US space official said:

> 'It's incredible, it would be contrary to the laws of science for a spacecraft in orbit to decay that fast unless there were some means of propulsion aboard to change the orbit.'

Have spacemen been taking more than just a passive interest in our satellites?

Having finished the astronomical section of this final chapter, let us now look at some interesting discoveries, which evidently are linked with ancient spacemen visits to Earth.

(A) A 1527 drawing of a rocket containing men, depicting what was supposedly seen in Europe that year.

(B) 2,000-year-old synthetic plastic excavated at an ancient Parthian site.

(C) Strange pink rocks at 12 sites in Eastern Pennsylvania; some of these rocks 'ring like bells' and are a complete mystery to geologists, mineralogists and chemists.

(D) A 700-year-old giant stone astronomical computer, the Ha 'amonga-'a Maui on Tonga.

(E) Statues 3 m 60 cm tall of an unknown race, discovered in 1964 on the tiny island of New Hanover, off the north-west tip of New Ireland, Bismarck Archipelago, Papua/Guinea. According to native legends these statues were made by the unknown race who arrived in antiquity, made 'magic stone circles' and faced the statues towards the rising sun.

THE AERIAL HORSEMEN

Were 'flying carpets' UFOs? The Cologne Bible of 1478–80 reproduces a woodcut portraying the Judaic story of the 'aerial horsemen'. That is, some eight men on horseback are mounted on a flying carpet which is travelling over a castle surrounded by a moat.

The 'aerial horsemen' are described in the second book of the Maccabees (the Hasmoneans), which are apocryphal works dating from about 200 B.C. The Maccabees were a Jewish group who fought for independence under Syrian domination.

Strangely, Lithuanian legends speak of 'flying boats', while a report in the *Thüringian Chronicle* describes three aerial 'fires' which entered the Hörselberg in Germany:

> 'In 1398 at midday, there appeared suddenly, three great fires in the air, which presently ran together into one globe of flame, parted again, and finally sank into the Hörselberg.'

Were they UFOs? An almost identical report originated from Weimar, Thüringia, in 1535!

Homer, about 900–800 B.C., related that the Phaeacians returned Odysseus to his homeland on board an extremely fast ship which travelled under its own volition: on 'automatic pilot'?

CONCLUSION

Law and Public Order in Space by McDougal, Yasswell & Vlastic (Yale University Press, 1966) raises the question of how a person would be charged under law if he was to kill or harm a visiting space being. Dr. Graham Hughes, a Professor at New York University, stated in the book that:

> 'As things stand now, an alien person would not have protection from the law, and could, therefore, be shot or impressed into slavery without running the risk of legal complications.'

Have space beings been visiting Earth for thousands or millions of years? Extraordinary facts, which can only really be explained by accepting this probability, assume new importance. One ancient curiosity is that Earthmen of old knew that outer space is black! The expression 'pitch darkness' or 'kromeshnaya tma', has been known in Russia for a long, long time, but the word kromeshnaya means 'outer' or 'external'; so this expression really means that deep space is black. The Chinese of antiquity also knew that space is black!

Did ancient Lemurians, Muans, Atlanteans and spacemen teach sciences to some Earthmen? Two secrets possessed by the ancients, were, firstly: curious texts left by the alchemists, which US scientists used as a 'valuable guide' to the secrets of nuclear fission, while developing the first A-bomb, and secondly: the exact method of preserving Egyptian mummies, which still is not known.

In February 1969, two US professors discovered 11 mummified bodies – many thousands of years older than Egyptian mummies – in an Argentinian cave. Symbols of the Moon, Sun, and Venus were also discovered in the cave.

Are our efforts to travel to other planets in spaceships, due to racial memories of our original homes in the sky? Could this explain the wide diversity among racial types on Earth, such as the Bayan-Khara-Ula Hill people (East Asia) who stand only 1 m 25 cm tall, Chinese, Negroes, and Caucasians? Were the ancestors of Earth's now four diverse racial types brought from other planets, perhaps each globe possessing but one people or racial type? Is *this* why many Earth peoples cannot live together in harmony, tending to stay among their own kind?

Did some Earthlings *not* wish to live on this planet? Were some placed here against their wishes for possible misdeeds on other worlds? Why, for example, did the ancients construct the 'Tower of Babel' (remains are possibly Birs Nimrud, about 12 m tall) and attempt to climb to 'Heaven', for which they were punished by forces from above? Did they subconsciously wish to return to their home, or homes, in the sky?

The three remaining tiers of the supposed Tower of Babel are made from brick, which in patches, have been *scorched and fused into lumps*, like the brickwork remains at Jericho!

Did spacemen destroy the Tower of Babel with an atomic missile?

The appearance of Birs Nimrud definitely signifies attempted razing with a projectile or missile!

R. Buckminster Fuller, renowned philosopher and visionary, referred to future scientific discoveries in *Saturday Review*, 29th August, 1964, when he said:

> 'We will probably learn that Darwin was wrong and that man came to Earth from another planet . . .'

The obvious point arises. Who really possessed or possesses Earth? We could even proceed one step further and consider a possibility which cannot be avoided in this query; i.e. the singularly disagreeable idea of Earthlings once under the dominant hegemony of other-determined, inter-star intelligences. That is, spacemen overlords possibly once owning, or even still owning, Earthman. However, benign colonization of Earth for multitudinous reasons could have been the intrinsic motivation for conveying settlers to our sphere.

Was Earth and our star system CLAIMED billions of years ago by aliens? Men from Lyra? Ophiuchus? Canis Major? Ursa Major? And why did the dinosaurs perish in such a manner? Were Earth's landlords defending their property from galactic raiders?

BIBLIOGRAPHY

Adamski, George — 1956. *Inside The Space Ships.* Arco Publishers Ltd & Neville Spearman Ltd, London.

Adamski, George — 1961. *Flying Saucers Farewell.* Abelard-Schuman Ltd, NY.

Agrest, Matest — *Astronauts of Yore.* Progress Publishers, Moscow.

Allingham, Cedric — 1954. *Flying Saucer From Mars.* Frederick Muller Ltd, London.

Aston, W. G. — *Nihongi or Chronicles of Japan.* G. Allen & Unwin Ltd, London.

Baumann, Hans — 1963. *Gold and Gods of Peru.* Oxford University Press, London.

Bellamy, Hans, Schindler and Allen, Peter — 1959. *Great Idol of Tiahuanaco.* Faber & Faber Ltd, London.

Best, Elsdon — 1955. *The Astronomical Knowledge of the Maori,* R. E. Owen, Government Printer, Wellington, NZ.

Blakney, R. B. — 1955. Translator of: *The Way of Life – Lao Tzu.* Mentor Books, NY.

Blavatsky, Helena, Petrovna — 1952. *The Secret Doctrine.* The Theosophical University Press, Calif.

Budge, E. A. Wallis — 1898. Translator of *The Book of the Dead.* Kegan Paul, Trench, Trübner & Co Ltd, London.

Burland, Cottie — 1965. *North American Indian Mythology.* Paul Hamlyn, London.

Carreyett, Ray A. — 1954. *Physical Anthropology.* W. & G. Foyle Ltd, London.

Cerminara, Gina — 1965. *Many Mansions.* William Sloane Associates Inc, NY.

Churchward, Col. — 1959. *The Lost Continent of Mu.* Neville Spearman Ltd, London.

Clark, Kate, McCosh — 1896. *Maori Tales and Legends.* David Nutt, London.

Colbert, Edwin H. — 1961. *Dinosaurs: Their Discovery And Their World.* E. P. Dutton & Co Inc, NY.

Colbert, Edwin H. — 1968. *Men and Dinosaurs.* Evans Brothers Ltd, London.

Conrad, Jack — 1967. *The Many Worlds of Man.* Macmillan and Company Ltd, London.

Coon, Carelton S. — 1963. *The Origin of Races.* Jonathan Cape Ltd, London.

Cowan, James — 1930. *Fairy Folk Tales of The Maoris.* Whitcombe & Tombs Ltd, NZ.

Cramp, Leonard G. (MSIA) — 1954. *Space, Gravity and the Flying Saucer.* T. Werner Laurie Ltd, London.

Dobzhansky, Theodosius — 1956. *The Biological Basis of Human Freedom.* Columbia University Press, NY.

Donnelly, Ignatius — 1960. *Atlantis.* Billing & Sons Ltd, London.

Dorson, Richard M. — 1962. *Folk Legends of Japan.* Charles E. Tuttle Co, Tokyo.

Dupont-Sommer, A. — 1961. *The Essene Writings From Qumram.* Translator, G. Vermes. Basil Blackwell, Oxford University Press, London.

Eiseley, Loren — 1957. *The Immense Journey.* Random House, Inc, NY.

Elwin, Verrier — 1958. *Myths of the North-East Frontier of India.* S. N. Gu Ha Ray at Sree. Saraswaty Press Ltd, Calcutta.

Giles, J. A. (DCL) 1849. Editor of: *The Venerable Bede's Ecclesiastical History of England*. And *The Anglo-Saxon Chronicle*. Henry G. Bohn, London.

Hansen, L. Taylor 1963. *He Walked The Americas*. Neville Spearman Ltd, London.

Hansford, S. Howard 1968. *Chinese Carved Jades*. Faber & Faber Ltd, London.

Heindel, Max 1937. *The Rosicrucian Cosmo-Conception*. Oceanside, Calif. Rosicrucian Fellowship.

Hodges, E. Richmond 1876. *Cory's Ancient Fragments*. Reeves & Turner Ltd, London.

Hodson, Geoffrey 1955. Lecture Notes. *The School of the Wisdom. Vol. 2.* The Theosophical Publishing House, Adyar, Madras.

Howell, F. Clark, and the editors of *Life* 1966. *Early Man*. Time-Life International. Nederland, NV.

Jacolliot, Louis 1874. *Histoire des Vierges: Les Peuples et les Continents Disparus*, Paris.

Josephus, Flavius 1889. *Antiquities of the Jews*. (Whiston's translation), George Bell & Sons Ltd.

Kazantsev, Aleksandr *A Visitor From Outer Space*. Foreign Languages Publishing House, Moscow.

Kofman, Janna 1967. *Almastis Exist*. Progress Publishers, Moscow.

Komarov, Vladimir 1967. *Why Dinosaurs Perished*. Progress Publishers, Moscow.

Leslie, Desmond and Adamski, George 1954. *Flying Saucers Have Landed*. Book I by Desmond Leslie.

Lewis, Lionel Smithett (MA) 1955. *St. Joseph of Arimathea at Glastonbury*. James Clark & Co Ltd, London.

Marsh, Frank Lewis	1963. *Evolution or Special Creation?* Review and Herald, Washington, DC.
McGrady, Patrick M. (Jr)	1971. *The Youth Doctors*. Ace Books Inc, NY.
Miller, R. DeWitt	1947. *Impossible, Yet it Happened*. Ace Books Inc, NY.
Murchie, Guy	1962. *Music of the Spheres*. Martin Secker & Warburg, London.
Pauwels, Louis, and Bergier, Jacques	1963. *The Dawn of Magic*. Anthony Gibbs & Phillips Ltd, London.
Poignant, Roslyn	1967. *Oceanic Mythology*. Paul Hamlyn, London.
Porshnev, Prof. Boris	1967. *Riddle of the Caucasian Shaitans – Is Neanderthal Man Extant?* Progress Publishers, Moscow.
Reed, A. W.	1961. *Myths and Legends of Maoriland*. A. H. & A. W. Reed, Wellington, NZ.
Rose, William, Stewart	1915. Translator of: *The Orlando Furioso of Ludovico Ariosto*. George Bell & Sons Ltd, London.
Scully, Frank	1950. *Behind The Flying Saucers*. Victor Gollancz Ltd, London.
Seltman, Charles	1961. *The Twelve Olympians*. Pan Books Ltd, London.
Shepherd, Walter	1961. *Our Universe*. Ward Lock & Co Ltd, London.
Smith, Richard Gordon (FRGS)	1908. *Ancient Tales and Folklore of Japan*. A. & G. Black Ltd, London.
Swift, Jonathan	1909. *Gulliver's Travels*. J. M. Dent & Co, London.

Trench, Brinsley le Poer — 1960. *The Sky People*. Neville Spearman Ltd, London.

Verrill, Prof. A. Hyatt — 1950. *The Bridge of Light*. Fantasy Press, NY.

Vyāsa The Compiler — 1956. *The Mahābhārata of Vyāsa Krishna Dwaipayana*. The Janus Press, London.

Waters, Frank — 1969. *Pumpkin Seed Point*. Sage Books, NY.

Whitelock, Dorothy — 1961. Editor of: *The Anglo-Saxon Chronicle* – Revised Translation, Eyre & Spottiswoode Ltd, London.

Wilkins, Harold T. — 1967. *Flying Saucers on The Attack*. Ace Books Inc, NY.

Williams, Jay — 1964. *Joan of Arc*. A Cassell Caravel Book, London.

Williamson, George, Hunt — 1965. *Road in the Sky*. Neville Spearman Ltd, London.

Winchester, A. M. — 1964. *Biology and its Relation to Mankind*. Van Nostrand Woltereck, H, NY.

Wiseman, P. J. — 1948. *New Discoveries in Babylonia about Genesis*. Marshall, Morgan & Scott, NY.

Yogananda, Paramahansa — 1968. *Autobiography of a Yogi*. Self-Realization Fellowship Publishers, LA, Calif.

MISCELLANEOUS

1848. *Six Old English Chronicles*. Henry G. Bohn, London.

1971. *Encyclopaedia Britannica*. William Benton Publishers, Chicago.

1967. *Everyman's Encyclopaedia*. J. M. Dent & Sons Ltd, London.

1907. *The Holy Bible*. King James Translation. Henry Frowde, Oxford University Press, London.

The Apocryphal New Testament. William Reeves, London.

1962. *India – Madhya Pradesh*. Government Publication, New Delhi.

MAGAZINES & NEWSPAPERS

Agrest, Matest — *Australian Flying Saucer Review* 1961, Feb, Vol II, No. 1. Epsilon Eridani – Earth.

Ahearn, Anthony — *Saga*. 1969, Sep. Mankind, Children of the Planets?

Beeson, Irene — *The New Zealand Herald*. 1968, Oct 10. Science May Soon Show Whether Pyramid Holds Ancient Treasure.

Belitzky, Boris — *Soviet News*. 1971, Oct 27. Life on Mars?

Berrill, N. J. — *The Atlantic Monthly*. 1957, Jun. Article about missing planet between Mars and Jupiter.

Carter, George F. — *Science Digest*. 1957, May. Mystery of Indian Civilization.

Cheetham, Robert N. — *Fate*. 1964, Apr. Bear Paws or The Feet of Buddha?

Cohen, Daniel — *Science Digest*. 1965, Dec. The Astronomy Story.

Cornish II, Joseph J. — *Science Digest*. 1956, Sep. Mystery of the Boomerang.

Duggan, Michael G. — *UFO Bulletin* (UFOIC, Sydney), 1958, Jun. Radio Communication from Outer Space.

Duggan, Michael G. — *Australian Flying Saucer Review* 1962, Jan, No. 6. Cosmic Intervention.

Evans, Gordon H. — *Fate*. 1964, Sep. Three Martian Mysteries.

Frank, Pat — *Saga*. 1963, Apr. Murder in Outer Space.

Gentet, Robert E. — *The Plain Truth*. 1971, Jan. Living Fossils.

Gerasimov, Gennady — *Soviet News*. 1971, Feb 12. Don't Blame Biology For Your Aggression.

Goodavage, Joseph — *Flying Saucers*. 1967, No. 3. Technology: 1,000,000 B.C.

Goodavage, Joseph — *Flying Saucers*. 1967, No. 4. Super Scientists From Nowhere.

Grabbe, Lester — *The Plain Truth*. 1970, Aug–Sep. Origin of Languages.

Heerebrand, P. W. — *People*. 1964, May 6. Was The Lost Paradise in Australia?

Hinfelaar, Henk J. and Brenda — *New Zealand Scientific Space Research* (*NZSSR*) Newsletter Nos. 22, 32. Journal (Spaceview) Nos. 35, 45, 49, 52, 58. Publication issued by Henk & Brenda Hinfelaar.

Jenness, Diamond — *The Corn Goddess* – Tales From Indian Canada. 1956, Bulletin No. 141, Anthropological series No. 39.

Kaust, Milton — *Rosicrucian Digest*. 1959, May. When The Earth Died.

Kazantsev, Aleksandr — *Australian Flying Saucer Review*. 1962, Jan, No. 6. Space Visitors.

King, Michael — *Weekly News* (NZ) 1969, Nov 11. Ancient Carving Said To Have Strange Powers.

Kroll, Paul — *The Plain Truth*. 1970, Jan. The Day The Dinosaurs Died.

Kroll, Paul and Hughes, Gene R. — *The Plain Truth*. 1970, Jun–July. Missing Link . . . Found!

Krylov, Igor (MSc) — *Znaniye-Sila*. 'Stone Brain' was Debunked, but . . .

La Farge, Oliver — *Science Digest*. 1956, Jan. Unsolved Mystery of the Mayas.

Long, E. John — *Science Digest*. 1956, Jun. Hunting Big Heads in Mexico.

Madden, Ross — *The Auckland Star*. 1967, Feb 25. Mystery of a Thousand Lost Men.

Madden, Ross — *The Auckland Star*. 1971, Feb 25. Atlantis At Last?

Marx, Robert F. with Rebikoff, Dmitri — *Argosy*. 1969, Dec. Atlantis At Last?

Marx, Robert — *Argosy*. 1971, Nov. Atlantis: The Legend is Becoming Fact.

McCoy, Robert B. — *Science Digest*. 1958, Oct. Mystery of the Medicine Wheel.

McIlraith, Shaun — *People*. 1964, Mar 11. Glass From The Moon.

Moser, William E. — *Australian Flying Saucer Review*. 1966, Nov, No. 9. Trans-Plutonian Planet.

Rampa, Lobsang — *People*. 1960, Aug 17. Excerpts from Doctor From Lhasa.

Reinart, Jeanne — *Science Digest*. 1966, Sep. The Picture That Hangs in Mid-air.

Robinson, Dr James — *Los Angeles Times*. 1966, Jul. 13. Article on Coptic manuscripts.

Sanderson, Ivan T. — *Argosy*. 1968, Feb. First photos of 'Bigfoot', California's Legendary 'Abominable Snowman'.

Schwartz, Stephan — *Coronet*. 1969, Feb. Is This The Year Atlantis Will Rise Again?

Sharland, Michael (FRZS) — *People*. 1965, Dec 29. Nature The Sculptor.

Sprague, Wallace A.	*Science Digest*. 1952, May. Mystery of the Green Fireballs.
Terry, Michael (FRGS, FRGSA)	*People*. 1963, Jan 16. Mystery Carvings in the Centre.
Terry, Michael (FRGS, FRGSA)	*People*. 1968, Jun 5. Were These The First Australians?
Tomas, Andrew	*Australian Flying Saucer Review*. 1961, Jul, No. 5. Skyships of Old.
Tomas, Andrew	*Australian Flying Saucer Review*. 1962, Nov, No. 7. Mt Shasta Mystery.
Tomas, Andrew	*Australian Flying Saucer Review*. 1965, Jun, No. 8. Phaethon and Photons.
Tunstall, John	*The Times*. 1969, Jul 26. Article on Chephren Pyramid.
Zaitsev, Vyacheslav	*Soviet Union*. 1968. No 12. Spacemen or Angels.
Zaitsev, Vyacheslav	*Sputnik*. Nos. 1 & 2. Visitors From Outer Space.

MISCELLANEOUS

AMORC booklet. 1943, May. The Hidden Archive.
Argentina. (booklet). Sawyers Inc. Ore.
Awake. 1969, Aug 22, and 1970, Oct 8.
Link. 1972, Apr. No. 30.
Man Jr. 1970, Jun. Mystery of 50 corpses.
New Scientist. 1972, Jun 29.
NZ Woman's Weekly. 1965. Space Travel article.
People. Astrology article.
People. 1962, Jan 3. Radioactive Vaults.
Rosicrucian Digest. 1959, Oct and 1969, Jan.
Science News Letter. Nos. Jun 18, 1960; Nov 26, 1960; Aug 26, 1961; Jul 7, 1962; Mar 30, 1963; Jan 11, 1964; Aug 21, 1965; Nov 22, 1969; Feb 13, 1971.
Science News. Nos. Aug 3, 1968; Apr 26, 1969; Sep 6, 1969; Jan 1970. Apr 14, 1973.

Spacelink. 1968, Mar Vol. 5, No. 2. Signs From Heaven. 1970, Jan Vol. 6. No 2.
Sunrise. Oct 1969.
The Humanoids. Special Issue. FSR. Oct–Nov, 1966.
The Path of the Soul. 1964. Booklet published in Amritsar.
Theosophy. 1966, Nov and 1970, Oct.
News items: *Soviet News. The New Zealand Herald. The Auckland Star*.
Acknowledgements to any sources of data, etc., which may have inadvertently been omitted.